PROBLEMS FOR COMPUTER SOLUTION

PROBLEMS FOR COMPUTER SOLUTION

SECOND EDITION

DONALD D. SPENCER

HAYDEN BOOK COMPANY, INC.
Rochelle Park, New Jersey

Library of Congress Cataloging in Publication Data

Spencer, Donald D
Problems for computer solution.

1. Programming (Electronic computers)—Problems, exercises, etc. 2. Problem solving—Data processing.
I. Title
QA76.6.S67 1979 001.6'425 79-20594
ISBN 0-8104-5191-3

Printed in the United States of America

2 3 4 5 6 7 8 9 PRINTING

81 82 83 84 85 86 87 YEAR

PREFACE

The purpose of this book is to bring together in one place a wide selection of problems to supplement the various programming language textbooks. Quite often programming texts do not have sufficient problems to give beginning students the practice they need.

PROBLEMS FOR COMPUTER SOLUTION was written specifically to serve as a supplement to any programming language text (BASIC, FORTRAN, APL, PL/1, etc.). The entire book is independent of any particular programming language.

The problems presented in this book are arranged by topic. Problems have been included from most mathematical disciplines (algebra, geometry, trigonometry, advanced mathematics, probability, statistics, number theory), science, chemistry, physics, business, biology, and game playing.

Within each chapter, an attempt has been made to arrange the problems in order of difficulty; however, since the difficulty of any problem depends upon the background and aptitude of the person attempting to solve it, the arrangement should be considered with reservations. Therefore, one should *not* assume that difficulty with a particular problem will imply greater difficulty with succeeding problems. Some problems include diagrams to aid the reader.

The book can be used either by students or teachers. I believe that programming cannot be learned by simply reading a description of how it should be done, but rather is learned by experimenting and doing. The wide range of problems will provide ample material for student exercises and will stimulate an interest in computing. A teacher may, of course, use the book as a "source book" for exercises, test material, problem assignments, etc.

One word of advice to students using the book. Be sure you fully understand the problem before attempting to solve it. Then develop an algorithm and draw a flowchart. Only then are you ready to write a computer program to solve the problem.

Happy problem solving!

Donald D. Spencer

CONTENTS

PROBLEMS FOR COMPUTER SOLUTION

CHAPTER 1

INTRODUCTORY PROBLEMS

This chapter contains a smorgasbord of introductory problems suitable for computer solution. These problems are ideal for "first problems," or for students just learning to solve problems with a computer. Many of the problems are little more than exercises.

1. Print the integers from 9 to 43.
2. Print the odd integers from 7 to 51.
3. Print the even integers from 2 to 48.
4. Print the integers from 1 to 30 paired with their reciprocals.
5. Print a table of powers of 2 which are less than 1000.
6. Convert inches to yards and feet and inches.
7. Determine if a given number is divisible by 14.
8. Determine if a given integer is a multiple of 6.
9. Input an integer and determine if it is "odd" or "even".
10. Write a program that accepts 25 positive integers as data and describes each as "odd" or "even".
11. Input a set of 25 numbers. Determine the number of positive numbers and the number of negative numbers in the set.

12. Find the largest even integer in a list of fifteen integers, some of which may be odd.

13. Determine the second largest integer in a set of 15 positive integers supplied as data.

14. Determine which is the greater quantity: 3^{75} or 2^{100}.

15. Print the addition table up to 12 + 12.

16. Print the multiplication table up to 12 × 12.

17. Compute and print the sum of the integers from 1 to 20.

18. Find the sum of 35 given integers.

19. Find the sum of all integers from 1 to 1000.

20. Find the sum of all the even numbers from 2 to 2000 inclusive.

21. Given any two numbers, find the sum and indicate whether the sum is positive, negative, or zero.

22. Input a positive integer N. Find the sum of the N integers 1, 2, 3, ..., N. Print each of the integers and the sum.

23. Input N numbers. Compute and print the product of the even numbers.

24. Calculate and print a two column table showing, in the first column, the integers 1 to n, and in the second column n^2. Do not use an integer greater than 30 for n.

25. The following table shows the fourth powers of the numbers 1 through 5. Compute and print a similar table containing the fourth powers of the first fifty integers.

NUMBER	FOURTH POWER
1	1
2	16
3	81
4	256
5	625

26. Print a table of square roots of the numbers 100, 101, 102, ..., 120.

27. Find the sum of the squares of the integers from 1 to N. That is, your program is to compute $1^2 + 2^2 + 3^2 + ... + N^2$.

28. Calculate the sum of the square roots of the "odd" numbers between 1 and 1000.

29. Convert dollars in decimal form to the equivalent in coins.

30. Convert P English pounds to D dollars and C cents. Use the exchange rate $2.80 = 1 pound.

31. Convert P English pounds, S shillings and E pence to D dollars and C cents. (1 pound = 20 shillings, 1 shilling = 12 pence).

32. Determine if a given integer is divisible by both 2 and 5.

33. Have the computer print the sum and the product of all possible different pairs of integers from 15 to 20.

34. Determine if the sum of $3^{1974} + 3^{1974} + 3^{1974}$ equals 3^{1975}

35. Write a program which asks for the constants W, X, Y and Z in the equation $WX + Y = YZ - Z$, and then prints the solution.

36. Calculate z according to the following formula. Assign the proper numeric variables. Let a = 4, b = 6, x = 8, y = 2, and c = 5 when you "run" your program.
    $z = (a + b)^3 - (x + y)^2 (a - c)^4 + 1/2(c + x)$

37. Find the largest of N nonzero numbers. Your program should calculate N by counting the number of nonzero values which precede a final zero.

38. Input 12 values of A and 10 values of B. Compute the sum of the A-values, the sum of the B-values, and the sum of the products AB.

39. Given ten integers, print only the largest. Do not assume that the numbers are listed in the data in any special order. You may assume that no two numbers are equal.

40. One inch is equivalent to 2.54 centimeters. Calculate the number of centimeters in 32 inches.

41. A formula to change kilograms to pounds is P = 2.2 × K where P = pounds and K = kilograms. Calculate the number of pounds in 212 kilograms.

42. Convert several weights in pounds and ounces into weights in kilograms.

43. If a printing press is moving paper through at a speed of 1000 feet per minute, what is the paper speed in centimeters per second?

44. If a certain grade of carpet has been selling for $9 per square yard, what should the price be per square meter?

45. The speed of light is represented as 2.99776×10^8 meters per second. Compute the number of meters in one light year.

46. The mean density of the earth is 5.522 g/cm^3. Determine the mass of the earth in grams.

47. A picture frame is 8 centimeters longer than twice its width (w). Express the length of the frame in centimeters.

48. Write a program which can convert the weights of the school football-team members from pounds to kilograms. (Hint: 1 kilogram = 2.204623 pounds).

49. Mariner 9 took 167 days to travel from Earth to Mars, a distance of about 34 900 000 miles. Express this distance in kilometers. What was its average speed in kilometers per hour?

50. In Miami, Florida, Lulu Rocket appeared at several social events in the role of "Miss Metric." Her "vital statistics" are: 89-58-89 centimeters. She is 170 centimeters tall and weighs 53 kilograms. Print her statistics in inches, her height in feet and inches, and her weight in pounds.

51. Compute and print the number of seconds in one week. In three weeks. In one month. In one month and three days.

52. Compute and print the number of seconds in D days, H hours, M minutes, and S seconds. For example, in 4 days, 6 hours, 24 minutes, and 11 seconds, there are 368 651 seconds.

53. Find and print the sum:
    1 + 3 + 5
    1 + 3 + 5 + 7
    1 + 3 + 5 + 7 + 9
    1 + 3 + 5 + 7 + 9 + 11

$1 + 3 + 5 + 7 + 9 + 11 + 13$
$1 + 3 + 5 + 7 + 9 + 11 + 13 + \ldots + (N - 1) + N$

54. Read N and a list of N numbers. Print these numbers in order of increasing magnitude.

55. Assume that everyone sleeps about 1/3 of the time (8 hours out of 24). Calculate how many hours a person has slept in a lifetime. A year has 365 days (disregarding leap years).

56. You have just been given a list containing the final exam grades for a class of 20 typing students. Count the grades below 65 and print out this number.

57. Find the average of N numbers. The value of N is to be input first followed by the N numbers.

58. To reverse a number is to write the number backwards (for example, the reverse of 123456 is 654321). Input any six digit number and find its reverse. Print the result in the following form:

    THE REVERSE OF 123456 IS 654321.

59. Find the absolute value of a number. If the number is zero, print ZERO. If the number is not zero, print the absolute value. The program should use the following integers as test data: 26, –400, 0, 216, –34.

60. Print the absolute value of -6, 0, 25, –143, –42.

61. Nancy took four tests. Her marks were 95, 68, 92, and 88. What was her average score?

62. Find the mean of the following 10 numbers: 75, 88, 84, 70, 65, 99, 91, 76, 43, 69. The program should sum the numbers and divide by 10 to obtain the arithmetic mean or average of the numbers.

63. Reorder a list of twenty integers by interchanging the first ten integers with the last ten integers.

64. Calculate n factorial, $n! = 1 \times 2 \times 3 \ldots (n - 1) \times n$, where n is any integer from 1 to 7.

65. Write a program that accepts three positive integers X, Y, and Z as data and computes the number X! + Y! + Z!.

66. Find the value of $x^4 - 8x^2 - 14x + 7$ for x = 2, 4, 6, 8, ..., 40.

67. Input a positive integer N followed by N social security numbers (SSN's). Each SSN is to be input in three parts: a three-digit number followed by a two-digit number followed by a four-digit number. Print the list SSN's as nine-digit numbers.

68. Input a positive integer, $N < 75$, followed by N real numbers. Print each of the numbers and the average of these numbers, then print all numbers which are within a five unit range of the average.

69. In the song "The 12 Days of Christmas," gifts are bestowed upon the singer in the following pattern: the first day she received a partridge in a pear tree; the second day, two turtle doves and a partridge in a pear tree; the third day, three French hens, two turtle doves, and a partridge in a pear tree. This continues for 12 days. On the 12th day she received 12 + 11 + ... + 2 + 1 gifts. How many gifts were there all together?

70. Arrange a set of three numbers in descending order. For example, for the data values 12, –7, 32, print 32 12 –7.

71. Write a program that prints 40 randomly selected integers K where $-100 < K < 100$.

72. Generate 20 sets of 50 random numbers, each having a value between 1 and 55. Print the largest random number obtained in each set of 10.

73. Generate X two-digit random numbers, and print all numbers which are less than your age, where X and your age are inputs.

74. Generate twenty random numbers in the range 1 - 100. Print the numbers as they were generated, and then print them in the following order: largest number, smallest number, second largest number second smallest number, and so on until the ten pairs of numbers have all been printed.

75. Print the pattern:

    1 0 0 1 0 1
    1 0 0 0 0 1
    0 1 1 1 0 1

    Then, change the dimension from 3 by 6 to 6 by 3 and print again.

CHAPTER 2

ALGEBRA

Programming a problem encourages a student to understand what he is doing, rather than rely upon reading similar sample problems. Further, the computer, by performing the arithmetic calculations, allows the student to concentrate on the problem or on the method of solution without wasting too much time.

The computer should not dominate or dictate the curriculum; rather, the computer should serve as an instructional aid in attaining the existing goals and objectives upon which a modern mathematics program is built. I have therefore tried to select problems that are normally stressed in elementary and intermediate algebra courses.

In this chapter you will find problems related to inequalities, word sentences, powers and roots, functions, graphs, systems of linear equations, polynomials, quadratic equations, irrational numbers, exponents, circular functions, complex numbers, exponential and logarithmetic functions, sequences and series, linear programming, and other areas often found in algebra courses.

1. Given an inequality $AX + B > C$ (A, B and C are real numbers), solve for X. For example, if the inequality is $4X + 14 > 34$, print $X > 5$.

2. Find the solution set of any inequality of the form $ax^2 + bx + c < 0$ for any values of a, b, and c. Test your program with the following inequalities:
$x^2 + 12x + 35 < 0$
$x^2 + x + 3 < 0$ $-x + 3x + 2 < 0$

3. Evaluate the rational expression
$$(2ab + 3b^2 + b)/(a^2b^3 - 368)$$
for a = 5 and b = 12. The program should print the answer in both fractional and decimal form.

4. Input a positive integer N and print the product P of the four consecutive integers N, N + 1, N + 2, and N + 3. P + 1 will be a perfect square.

5. Find the arithmetic mean of the numbers 60 and 68.

6. Find the square roots of the integers from 9 to 25. Print the integer and its square root.

7. Input two integers and without actually multiplying the numbers, determine whether their product is positive, negative, or zero.

8. Input a real number N. Print N, its additive inverse, and (if it has one) its multiplicative inverse.

9. Compute the square, cube, square root, and cube root of integers from 1 to 100. Print the results in table form.

10. Print a table of values for $y = a^x$.

11. Print a table of square, cube, and fourth roots of the first twenty integers.

12. Find the solution for the exponential equation $A^X = B$, where A = 3, and B = 81.

13. Given values for constants a, b, c, and d, and values of variable x, which are to be input to the program, write a program which will evaluate the function defined by
$$f(x) = \begin{cases} ax^2 + bx + c & \text{if } x < d \\ 0 & \text{if } x = d \\ -ax^2 + bx - c & \text{if } x > d \end{cases}$$

14. For each of the following pairs of numbers, find the greatest common factor: 60, 12; 35, 10; 28, 32; 65, 179; 210, 1036.

15. Have the computer generate pairs of integers and find the greatest common factor.

16. Find the greatest common factor of a given set of three numbers.
17. Have the computer generate sets of three integers and find the greatest common factor.
18. For each of the following pairs of numbers, find the least common multiple: 25, 645; 132, 360; 192, 24.
19. Generate pairs of integers and find the least common multiple.
20. Find the least common multiple of five given numbers.
21. Factor the following trinomials.
    (a) $6x^2 + 11x + 3$
    (b) $5x^2 + 31x + 6$
    (c) $10x^2 + 6x - 24$
    (d) $2x^2 - 41x - 336$
22. Factor the trinomial $30x^2 + 4x - 48$ into prime factors.
23. Given the number of a baseball team's wins and losses, compute its winning percentage. Assume that there are no ties.
24. Assume that a babysitter works for a rate of 75 cents per hour until 10 p.m. and 1 dollar per hour thereafter. Given the time the sitter's beginning and ending times, compute the charge for an evening's work.
25. The Chamber of Commerce of your city is sponsoring a number lottery with tickets numbered from 1 to 1500. Input a ticket number, T, and check the list of 12 lucky numbers to see if it is a winning ticket. Use random numbers to select the 12 winning numbers.
26. Steven has some comic books. He has exactly 3 times as many as Michael and exactly 4 times as many as Laura. They have less than 200 comic books among them. How many comic books could each one have? Show all possibilities.
27. Tomatoes cost 25¢ per pound more than potatoes. If potatoes cost x cents per pound, express the cost of 6 pounds of potatoes and 3 pounds of tomatoes.
28. A farmer with $100 goes to market to buy 100 head of stock. Prices are as follows: calves, $10 each; pigs, $3

each; chickens, $.50 each. He gets 100 head for his $100. How many of each does he buy?

29. Find the rate at which a person must travel to overtake another person that left the same place some time before.

30. Find the time it takes one person to overtake another if they leave at different times, travel at different rates, and travel in the same direction.

31. A rectangular swimming pool has dimensions 40 meters by 13 meters. A person standing at a corner A wishes to go to corner C in the least possible time. He has the option of running the entire distance around, swimming diagonally across, or combining some running with some swimming. He determines that the fastest route is the combination. His running rate is 1 meter/sec. and his swimming rate is .5 meter/sec. Find the running distance and the swimming distance which requires the least time to go from corner A to corner C. The program should output the running distance, the swimming distance, and the minimum time.

32. Solve the following problem: A boat travels at a rate of 6 miles per hour in still water. If it takes 4 hours to travel 12 miles upstream, find the rate of the river current.

33. Find the distance between two people who start at the same place at the same time and travel in opposite directions for a period of time at different rates.

34. If five pairs of birds each raise 3 eggs to adulthood, then die, leaving the remaining 15 birds to mate and also raise 3 eggs per pair to adulthood, then die, etc., how many birds will there be at the end of 5 years?

35. Bob can do 50% more work than Bill and 25% more work than Sam. Working together the three men need 15 days to build a swimming pool. Find the time needed by each man to do the job alone.

36. The owner of a candle shop could not afford to give sales clerk a raise. So they agreed to the following plan, which the clerk suggested. "I'll work Monday through Saturday for three weeks. On the first day you

pay me a penny, on the second day two cents, and on the third day four cents. Each day pay me double the amount of the day before." Write a program to determine the total amount the clerk got during the three weeks.

37. Arrange the digits from 1 to 9 in order, using only addition and subtraction, to total 100. For example,

$$1 + 23 - 4 + 56 + 7 + 8 + 9 = 100$$
$$12 + 3 + 4 + 5 - 6 - 7 + 89 = 100$$

Print all possible combinations.

38. A truck when fully loaded can carry enough fuel to take it halfway across a barren desert. If the truck can return to the starting point as is necessary, write a program to determine the minimum amount of fuel required to take it all the way across. Assume that any amount of fuel can be taken from the truck at any point in the desert and cached and that this amount will remain undiminished until subsequently collected.

39. Compute the systolic blood pressure of persons whose ages are 25, 35, 47, 51.5, 60. Use the formula $P = 100 + \frac{1}{2}A$ where A represents age.

40. Compute the height (h), in t seconds, of a body thrown upward with starting speed r.

$$h = rt - 16t^2$$

In this example, $t = 2$ and $r = 32$.

41. If two cities are 80 kilometers apart and you can drive 90 kilometers per hour driving between them, how many minutes will it take to drive the distance between them?

42. An International Football League is starting in direct competition with the National Football League. You purchase a franchise in the newly formed association, and you are told that your profit (in $1000 units) can be projected for the next 8 years by the formula

$$p = t^3 - 5t^2 + 10t - 51$$

(p represents your profit, t the time in years). At $t = 0$, the time of the purchase of the franchise, $p = 51$. Your cost of the franchise, therefore, is $51,000, a negative profit indicating a loss. Determine your total profit (or loss) for the cumulative 8 years.

43. The star athlete at Midwest High School throws a football, a baseball, and a medicine ball up in the air. A quadratic function gives the heighth in meters of each throw with respect to time in seconds as follows:

    Football: $f(t) = -16t^2 + 43t + 8$
    Baseball: $b(t) = -16t^2 + 100t + 12.5$
    Medicine ball: $m(t) = -16t^2 + 1.5t + 2$

    Print the time t in seconds and the height of each ball after t seconds, where t is an integer between 0 and 10 inclusively.

44. Find the power of a power, i.e., find $(a^m)^n$, where $a = 20$, $m = 3$, $n = 2$.

45. Compute the values of 2^2, 2^3, 2^4, and 2^5; 3^2, 3^3, 3^4, and 3^5; and so on through 9^2, 9^3, 9^4, and 9^5.

46. Change the repeating decimal 2.181818 to a rational fraction of the form M/N where M and N are integers.

47. Test a given number to determine whether it is prime. Print PRIME if it is and NOT PRIME if it isn't.

48. Determine the absolute value of a number. Do not use the absolute value function.

49. Input several values for A and B to test the truth of the expression

$$|A + B| \overset{=}{+} |A| + |B|$$

50. Solve an absolute value equation of the form $|X - A| = B$, where A and B are real numbers.

51. Print the elements of the function defined by $f(x) = |x|$, for $x = -8, -7, -6, \ldots, +7, +8$.

52. Evaluate the function $y = \sqrt{x}$ as x takes on integral values from 1 to 10 inclusive.

53. Find the general form of the linear equation given the coordinates of two points on the line: (9,7) and (5,4).

54. Input real root R. Determine by substitution if R is a root of the quadratic equation $Ax^2 + Bx + C = 0$.

55. Input a real number X. If X is non-negative, print the principal fourth root of X. If X is negative, print the message NO REAL FOURTH ROOT.

56. Input A, B, and C (real numbers), which are the coefficients of a quadratic equation $Ax^2 + Bx + C$

= 0. Determine if the equation has real roots. Print one of the messages: REAL ROOTS or NO REAL ROOTS.

57. Input real numbers A, B, and C, which are the coefficients of a quadratic equation $AX^2 - BX + C = 0$. Determine if the equation has one or more real roots. If so, compute and print the root(s).

58. Given a point P(x,y), determine the ordered pair of the point which is symetric to P with respect to the x-axis.

59. Find where the line represented by the equation $y = 4x + 3$ crosses the x-axis.

60. Plot the curve $y = x^2$ from $x = -6$ to $x = +6$. Label the horizontal and vertical scales.

61. Plot the curve $y = 4x^2 - 5x + 2$ from $x = -3$ to $x = 5$. Label the horizontal and vertical scales.

62. Find the zeroes of the quadratic function $f(x) = x^2 - 4x - 165$.

63. Find the vertex for the quadratic function $f(x) = 3x^2 + 18x + 7$.

64. Evaluate the polynomial $f(x) = 12x^2 + 6x + 8$ as x takes on values from 1 to 10 in steps of 0.1.

65. Given a linear equation $AX + B = C$ (A, B, and C are real numbers), solve for X. For example, if the equation is $6X + 12 = 30$, print $X = 3$.

66. Write a program to evaluate the function defined by $y = 3x^2 + 4x - 1$ for x, where $-10 \leq x \leq 2$ and x is an integer.

67. Find all solutions of $12x - 18y + 4 = 0$ for $x = 5, 10, 15, ..., 40$.

68. Find all solutions of $5x + y + 17 = 0$ for $x = -8, -2, -1.5, 2, 5, 6, 12, 15$.

69. Find the vertex, axis of symmetry, and zeroes of the quadratic function $f(x) = 3x^2 + 5x - 2$.

70. Factor a polynomial of the form $x^2 + bx + c$.

71. Find the zeroes of the function $y = x - 1/3$.

72. Use the quadratic formula in your program to solve equation $15x^2 - 23x + 41 = 0$.

73. Solve the following equation: $6x^2 - 17x + 5 = 0$.

74. The two roots of a quadratic equation, $ax^2 + bx + c = 0$ can be found by the formula

$$\text{roots} = (-b \pm \sqrt{b^2 - 4ac})/2a$$

Write a program to find and print the roots for any given inputs. If $b^2 - 4ac$ is negative print NEGATIVE DISCRIMINANT.

75. Determine whether any integers from –10 through 10 solutions of $x^3 + 2x^2 + 75 = 0$.

76. Each of the following equations has two real solutions. Print a table containing the values of a, b, c, the two solutions, the sum of the solutions, and the product of the solutions.

(a) $x^2 - 3x - 54 = 0$
(b) $2x^2 + x - 3 = 0$
(c) $9x^2 + 45x - 18 = 0$
(d) $21x^2 + 11x - 2 = 0$

77. The graph of $4x - y - 6 = 0$ intersects the y-axis at a point (0,–6); this point is called the y-intercept of the line. The x-intercept is the point where the line intersects the x-axis; the y-coordinate of this point is, of course, 0. The x-intercept of this line is (1.5,0). Write a program to find the x-intercept and y-intercept of the line represented by the equation $6x - y + 43 = 0$.

78. Determine the x- and y-intercepts for the equation $4x + 3y = 6$.

79. Input the coordinates of a point and determine if the point lies on, above, or below the line $y = x$. The program should input the following point coordinates: (1,1); (3,4); (–3,–4); (–6,7); (–5,5); and (–1,3).

80. Read the coordinates of a point in the xy plane. Identify the quadrant in which the point lies or, if it lies on an axis, identify which axis.

81. Print the equations of lines parallel and perpendicular to the line represented by $6x - 18y = 36$ which pass through the point (–6,–2).

82. Determine the equation of the line that passes through the points (0,–2) and (–68,–15).

83. Determine the equation of the line that passes through the points (56,16) and (–40,1).

84. Determine the equation of the line described by the slope 3 and the point on the line (8,–4).

85. Print the slope of the line with a y-intercept of (0,10) and a point on the line (–3,0).

86. Find the slope, the x-intercept and the y-intercept of the graph of the equation $2x + 3y + 8 = 0$.

87. Determine the slope and y-intercept for each of the following lines:

    (a) $3x + y = 4$
    (b) $x - y = 2$
    (c) $5x - 3y = 15$

88. Compute the slope of the line passing through two given points. Use the following pairs of points as test data.

    (a) (–3,–5) and (0,–2)
    (b) (–4,6) and (8,–3)
    (c) (3,–5) and (3,0)

89. Find the slope of the line through two given points in the coordinate plane. Include the possibility that the slope may be undefined.

90. Write a program to input four numbers, print the additive inverse of each, the sum of the four numbers, and the additive inverse of the sum.

91. A lighthouse is located at coordinates (7.64,12.12). A boat initially located at (2.00,0.35) is moving in a linear direction which will take it past the lighthouse. After one minute, the location of the boat is (3.37,1.87). Determine the coordinates (x,y) on the boat's path when the distance from the boat to the lighthouse is a minimum (to 2 decimal places).

92. Input two pairs of coordinates. Find the slope and the y-intercept of the straight line containing the points and print the results as rational numbers reduced to lowest terms. If the result is negative, make the numerator the negative number.

93. Input two ordered pairs and print the equation of the perpendicular bisector of the line segment determined by these points.

94. Input A, B, and C for the parabola $y = Ax^2 + Bx + C$,

where $A \neq 0$. Compute the equation of the directrix, and the coordinates of the focus.

95. Write a program that graphs an equation of a straight line $y = mx + b$.

96. Write a program that computes the range for a given domain and then graphs the function. Produce a solution for the function $y = x^2 + 14x - 1$.

97. Write a program to graph a portion of the function defined by $y = .1x^2 - .2x$.

98. Find the equation of a line that is perpendicular bisector of a line segment whose end point coordinates are (2,3) and (2,9).

99. Find the directed distance from a line (represented by the equation $x + 2y + 3$) to the point (4,5).

100. Find the X and Y coordinates for the point which divides a line segment in a given ratio. For example, find the coordinates of the point which divides the line segment whose end points are (0,–2) and (3,7) in the ratio of 1 to 2.

101. The greatest integer function is denoted by $(fx) = [x]$. The function associates with each real number x the largest integer not greater than x. Find the value of the greatest integer function given an argument that is positive, negative, or zero.

102. Input positive integers A and B and determine the quotient and the remainder when A is divided by B.

103. Compute and print the determinant of a 2 × 2 matrix.

104. Evaluate the second order determinant

$$\begin{vmatrix} 7 & 8 \\ 42 & -73.3 \end{vmatrix}$$

105. Write a program that uses synthetic division to find the quotient and remainder when a polynomial of any degree is divided by a linear polynomial of the form $x - b$.

106. Write a program which will perform synthetic division.

107. You are given two equations of the form

$$4x + 5y = 17 \qquad 3x + 6y = 22$$

Find the values of x and y.

108. Input the coefficients for the system of linear equations: $Ax + By = C$ and $Dx + Ey = F$. Determine whether the graphs intersect. If they do, do they intersect in one point or in infinitely many points?

109. Read the coefficients for the system of linear equations: $Ax + By = C$ and $Dx + Ey = F$. Determine whether the graphs of the equations are perpendicular.

110. Input the coefficients for the system of linear equations: $Ax + By = C$ and $Dx + Ey = F$. Determine whether the equations are consistent or inconsistent, and dependent or independent.

111. Print whether or not the following linear equations have the same graph.

$$y = 7x + 9 \qquad 3y - 21x = 12$$

112. Print whether the lines determined by the equations

$$y = 5x + 19 \qquad y = -4x - 19$$

are perpendicular.

113. Print whether the graphs of the following two equations are the same line, parallel lines, or lines that intersect in one point.

$$5x + y = 12 \qquad 2y = -10x + 24$$

114. Solve the following system of equations.

$$3x + 4y - 6 = 0 \qquad 2x + 3y = 0$$

115. Solve the system of linear equations:

$$109x + 71y - 260 = 0$$
$$-89 + 29y + 18 = 0$$

116. Solve the following 3×3 system of equations:

$$3x - 5y - 2z - 5 = 0$$
$$-5x + 2y + 3z - 12 = 0$$
$$2x + y + 0z - 5 = 0$$

117. Solve the following system of linear equations:

$$2x + 4y - 5z - 3 = 0$$
$$3x - 2y - 2z + 14 = 0$$
$$-4x + 5y + 3z + 10 = 0$$

118. Write one program that will solve the three equations that follow.

$$x + 6y + 1 = 0$$

$$2x - y + 5 = 0$$
$$-43 + 13y - 6 = 0$$

Let $x = 2, 4, 6, 8, \ldots, 50$.

119. Find the center and radius of a circle given by the conic equation $x^2 + y^2 - 144 = 0$.

120. Find the center and radius of a circle given the conic equation $x^2 + y^2 + 2x + 4y - 20 = 0$.

121. Find the Nth term and the sum of the first N terms of an arithmetic progression. Use the formula $L = A + (N - 1)D$ to find the last term, and $S = (N/2)(A + L)$ to find the sum of the terms. In this example, $A = 1$, $N = 2$, and $D = 3$.

122. Find the sum of an arithmetic series $A + (A + D) + (A + 2D) + \ldots + (A + (N - 1)D)$, for a given value of A, D, and N.

123. Determine the sum of the series $1 + 2 + 4 + 8 + 16 + \ldots + 2^n$ for any positive integer value of n.

124. Find the sum of a geometric series $A + AR + AR^2 + \ldots + AR^{N-1}$, for a given value of A, R, and N.

125. Compute the sum of 120 terms in an arithmetic series. Use the formula $S = 1/2n(n + 1)$. In this problem, $n = 120$.

126. Input the 30 elements of array A. Change positions of the following elements: A(2) and A(16), A(5) and A(25), and A(26) and A(12). Print the array.

127. Compute the sum, difference, product, and quotient for pairs of complex numbers assigned as data.

CHAPTER 3

GEOMETRY

This chapter contains problems suitable for use in a typical course in geometry. By using the computer in the geometry classroom, students may compute areas and volumes of geometric figures with great accuracy, generate Pythagorean triples endlessly, and explore many areas which were previously inaccessible.

1. Given the three sides A, B, and C of a triangle, find the three angles a, b, c. Assume that all the angles are acute.
2. Given an angle measure greater than 0 and less than 180 degrees, classify the angle as obtuse, right, or acute.
3. Input D, the degrees in an acute angle, and compute the measure of its complement and supplement.
4. Determine the angle between two intersecting lines.
5. Input the measures of two remote interior angles of a triangle. Determine the measure of one of the external angles.
6. Input the measure of the vertex angle of isosceles triangle ABC with AB = AC, and determine the measure of a base angle.

7. The distance between two points A and B is defined as $|A - B|$. Write a program to compare $|A - B|$ and $|B - A|$. Use the following replacements for A and B for your test data.

A	12	-5	-2	9
B	10	9	-3	13

8. Find the area of any rectangle by the formula Area = lw, where l is the length and w is the width.

9. Find the third side of a right triangle using the pythagorean theorem.

10. Determine the length of the hypotenuse in each of these right triangles: ARB, BRC, CRD, DRE, ERF.

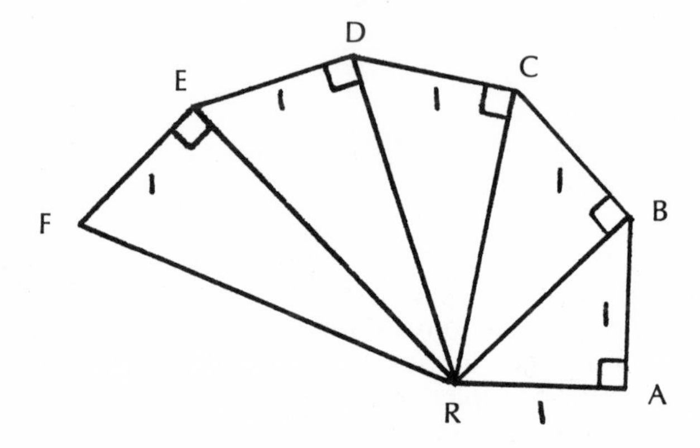

11. If a dress pattern calls for 3.5 yards of material 45 inches wide, how many meters will be required of material 110 centimeters wide?

12. The sum of the angles of a triangle is 180°. Input two angles A and B, and compute the value of the thrid angle, C. The program should check for a third value that is zero or negative, and if either value exists, print the message NOT A TRIANGLE.

13. A basic geometry theorem deals with the possible measures of the 3 sides of a triangle. The theorem states that the sum of the measures of the sides of a triangle must be so arranged that the sum of the measures of any two sides is greater than the measure of the third. Write a program which will determine if any three numbers can be the measures of the sides of a triangle.

14. Input three positive numbers X, Y, and Z. Determine if these could be the lengths of the sides of a right triangle.
15. Write a program which will find the three angles of a triangle given the three sides.
16. Given the three sides of any triangle ABC, compute and print the area of the triangle ABC.
17. Input the lengths of the sides of a triangle. Determine whether the triangle is Isosceles, Equilateral or Scalene.
18. Input the lengths of the two legs of a right triangle and compute the perimeter.
19. Input the lengths of the sides of a triangle. Find the perimeter.
20. Input the lengths of the three sides of a triangle, and determine the area.
21. Given any three parts of a triangle ABC, one of which must be a side, compute and print the area of triangle ABC.
22. Given two sides and the included angle of any triangle ABC, compute and print the area of triangle ABC.
23. Determine the perimeter of an isosceles right triangle, given the length of a leg.
24. Input the length of the hypotenuse of an isosceles right triangle and compute the length of a leg.
25. Input B, the base, and H, the height of a triangle, and determine the area.
26. Snoopy, a giant from another planet has decided to invade earth and then return to his own planet. During his visit, he plans to conceal his identity by wearing a mask over his nose and mouth. The mask must have a height (h) of 6.4 meters and a base (b) of 14.3 meters. Using the equation AREA = 1/2bh, write a program to determine how many square meters of material the giant will need.
27. The area of a right triangle is equal to twice its perimeter.

    $$1/2(A \times B) = 2(A + B + C)$$

    The sides of the triangle are integers each less than 100. Find the triangles.

28. Input X, the length of one side of an equilateral triangle, and compute the perimeter.

29. Determine the perimeter of a right triangle, given the lengths of the two legs.

30. Determine the perimeter of a right triangle, given the lengths of the hypotenuse and the length of one leg.

31. Input the lengths of three sides of a triangle and the lengths of the three corresponding sides of a second triangle. Determine whether the triangles are similar.

32. Input the lengths of three sides of a triangle and the lengths of the three corresponding sides of a second triangle. Determine whether the triangles are congruent.

33. The co-ordinates of the vertices of two triangles are supplied as data. Find whether one triangle lies completely inside the boundary of the other.

34. Find the area of a triangle given the co-ordinates of the three vertices.

35. Hero's formula may be used to find the area of any triangle, given the measures of the three sides. The formula is

$$\text{Area} = \sqrt{s(s - a)(s - b)(s - c)}$$

where $s = 1/2(a + b + c)$. Find the area of a triangle with sides equal to 6, 8, and 10 meters.

36. Use Hero's formula to find and print the triangles whose areas are integral and whose sides are consecutive integers less than 1000.

37. Find the area of any square by the formula $A = S^2$.

38. Given the length of a side of a square, compute the area.

39. Given the length of a side of a square, compute the perimeter.

40. Compute the surface area S for a rectangular solid with dimensions l, h, and w. In this problem let l = 10 meters, h = 4 meters, and w = 5.2 meters.

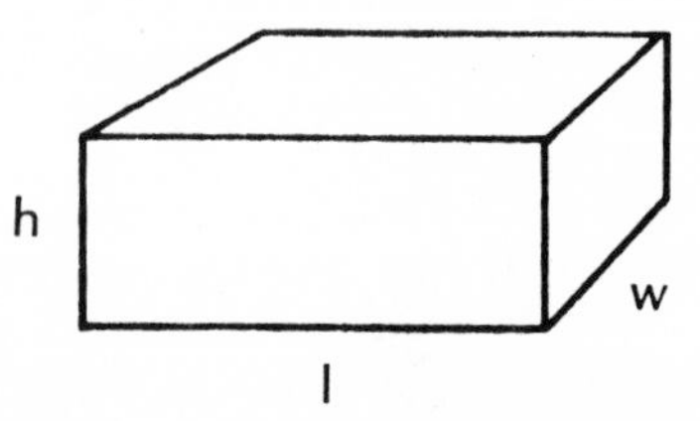

41. A wall 1 meter, 50 centimeters long and 1 meter, 10 centimeters high is to be covered with tile. Each piece of tile is 11 centimeters × 11 centimeters. What is the minimum number of pieces of tile needed?

42. If a living room is 10 feet × 15 feet, how much will it cost to carpet it, if carpeting costs $9 a square meter?

43. Will there be any change in the area of a rectangle if the length is doubled and the width is halved? Write a program to help you answer this question.

44. A room is to be wall-papered. It is 4 meters, 25 centimeters wide, 5 meters, 60 centimeters long, and 2 meters, 80 centimeters high. In the room there are two doors 1 meter wide and 2 meters, 30 centimeters high and one window 1 meter, 50 centimeters wide and 2 meters high. How many square meters of wallpaper will be needed? (Remember that a room has four walls and that no wallpaper is needed for the doors and the window).

45. Marlene decided to plant a rectangular watermelon bed. How much fence should she buy if the bed is to be 6 meters, 50 centimeters long and 4 meters, 70 centimeters wide?

46. A New Orleans singer lives in an efficiency apartment 6 meters long, 6 meters wide and 6 meters high. She wants to paint the walls and the ceiling with chartreuse paint that will cover 30 square meters per gallon. Write a program to determine how much paint she should buy.

47. The Medical Arts Building consists of 4 rooms with the following dimensions:

    Room 1 - 3.5 meters × 4.0 meters
    Room 2 - 4.5 meters × 5.5 meters
    Room 3 - 4.0 meters × 6.0 meters
    Room 4 - 5.0 meters × 8.5 meters

    Compute the floor space in this building.

48. A board measures 2 meters, 31 centimeters. If it is to be cut into 3 equal lengths, how many centimeters should each one be? If it is to be cut in the exact center, how many centimeters should be measured from one end?

49. A rectangle with length 6 meters and width 3 meters has an area of 18 square meters and a perimeter of 18 meters. Find another rectangle that has an area and perimeter of the same number.

50. A swimming pool is 7 meters wide, 14 meters long and has an average depth of 1.4 meters from the bottom to the surface of the water. What is the weight of the water in the pool in (a) kilograms, and (b) metric tons?

51. Compute and print the area and the perimeter of a parallelogram with input values of c = 8 meters, d = 4.2 meters, and h = 4 meters.

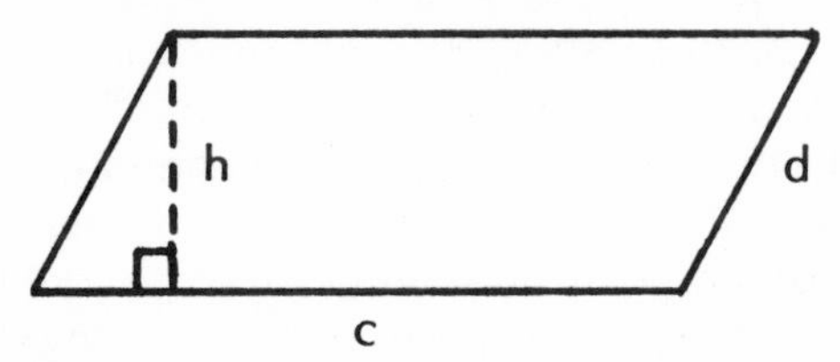

52. If a parallelogram has a base of 30 centimeters and a vertical height of 15 centimeters, what is its area?

53. The sides of a parallelogram are 35 meters and 50 meters, and the smaller angle is 20°. What is the length of the longer of the two diagonals?

54. Input the height and the length of the bases of a trapezoid. Determine the area of the trapezoid.

55. If a sphere has a radius of 8 centimeters, what is the area of its surface?

56. Input the radius of a sphere and compute its surface area.

57. Input the lengths of the four sides of a quadrilateral. Determine whether the quadrilateral is equilateral.

58. A kite has the shape of a quadrilateral with two pairs of adjacent sides congruent. The vertical diagonal is the perpendicular bisector of the horizontal diagonal. Find the area of the kite.

WY = 8
XZ = 12

59. Compute and print the volume of a cylinder of radius r and height h. In this problem, r = 10 centimeters, and h = 32 centimeters. Use the formula $V = \pi r^2 h$.

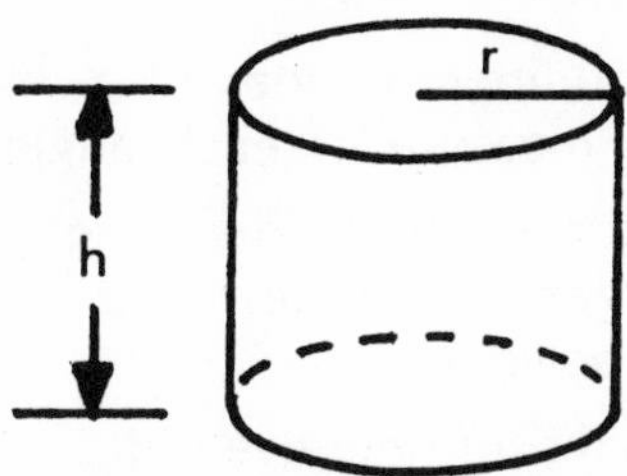

60. Find the volume of any cylinder when the radius of the base and the height of the cylinder are known.

61. Compute the surface area of a cylinder using the formula: $S = 2\pi(r^2 + h)$.

62. Compute the area of the paper wrapping on a can of corn that is 24 centimeters high and 12 centimeters in diameter. The formula to compute the outside area of a cylinder can be written

$$\text{Area} = \text{height} \times \text{diameter} \times \pi .$$

63. Input the length, L, of the altitude and the radius, R, of the base of a right circular cylinder. Determine the volume, the total surface area, and the lateral area of the cylinder.

64. Find the slant height of a regular square pyramid, given the length of a side of the base and the length of a lateral edge.

65. Compute the surface area of a regular pyramid where b is the length of each side of the base, and h is the altitude of each triangular face. In this problem, b = 10 and h = 17.

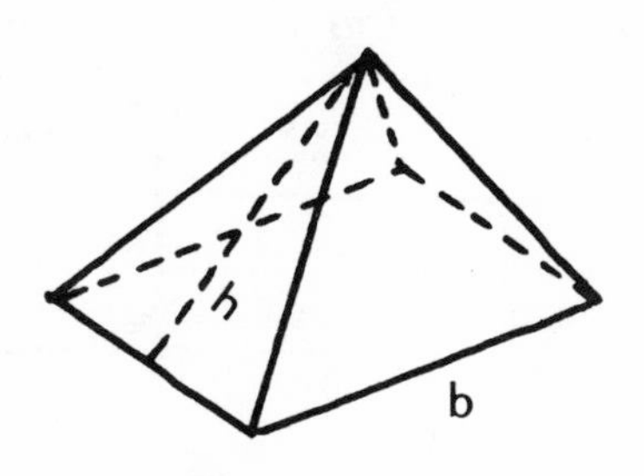

66. Input the length of a side of the base and the length of a lateral edge of a regular square pyramid. Determine the volume, surface area, lateral area, and slant height of the pyramid.

67. Input N, the number of sides of a regular polygon. Determine the measure of each angle.

68. Read X, a positive integer denoting the number of sides of a polygon. Compute the sum of the interior angles of the polygon.

69. Input the lengths of the five sides of a pentagon. Determine whether the pentagon is equilateral.

70. Compute the area of a polygon. Assume that the polygon has a known but variable number of sides.

71. Compute the area of a regular polygon, given the number of sides and the measures of the apothem and one side.

72. Compute and print the area A and length of perimeter P of a polygon with n sides circumscribed about a circle of radius r. Input values for r and n and output the value of A and P.

73. Calculate the volume and area of a sphere using the formulas $V = 4\pi r^3/3$ and $A = 4\pi r^2$ where r is the radius of the sphere. In this problem, r = 10 centimeters.

74. Input L, the length, W, the width, and H, the height of a rectangular prism. Compute the volume and total surface area of the prism.

75. Input the lengths of the diagonals and determine the area of a rhombus.

76. Compute the volume of a "plumb-bob" for input values of r, a, and b.

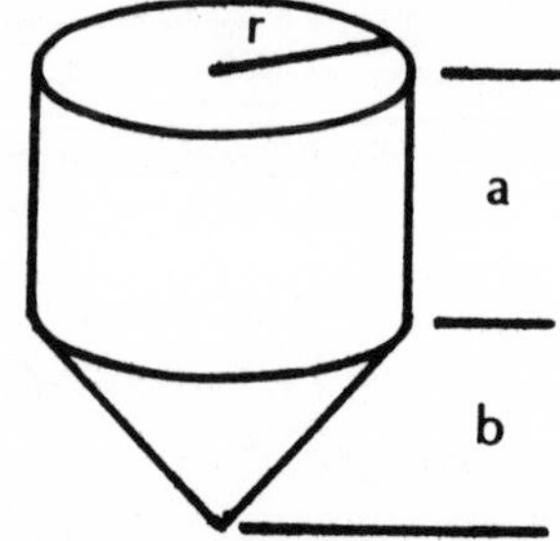

77. Input A, the altitude, and R, the radius of the base of a right circular cone. Determine the volume, lateral area and total surface area.

78. An engineer who builds earthfill dams wants a program to calculate the volume of earth required for a given dam. All dams are shaped as shown below; they vary only in dimension. Can you help this engineer by writing a program to calculate the volume in cubic yards?

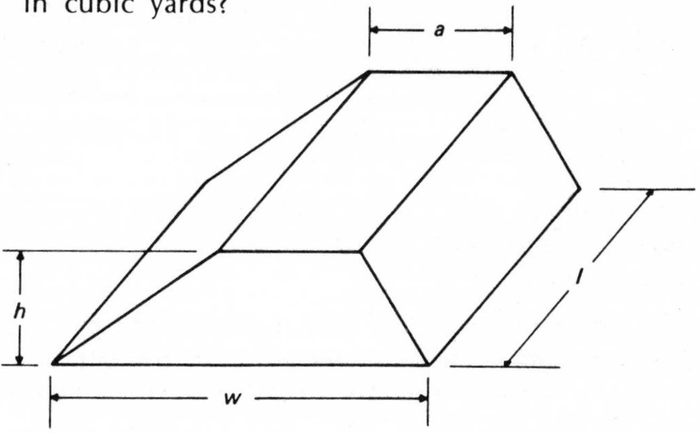

79. A Pythagorean Triple is a set of numbers which satisfy the relationship $A^2 + B^2 = C^2$. The numbers (3, 4, 5) and (5, 12, 13) are examples of Pythagorean Triples since $3^2 + 4^2 = 5^2$ and $5^2 + 12^2 = 13^2$. Find 15 triples of this type.

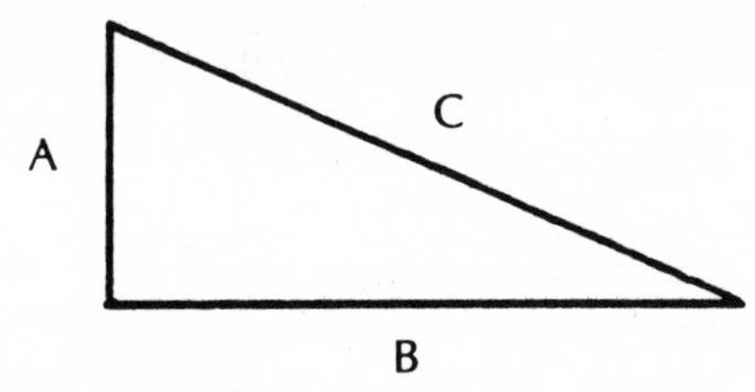

80. Find all Pythagorean triples with an hypotenuse less than or equal to 70.

81. Verify that the product of a Pythagorean triple is always divisible by 60.

82. Approximate π by taking the sum of the first twenty five terms in the formula $\pi/4 = (1 - 1/3 + 1/5 - 1/7 + 1/9 \ldots)$ and multiplying by four.

83. The distance, d, between any two points in the real plane can be determined by the formula $d = \sqrt{(x_2 - x_1)^2 + (y_2 - y_1)^2}$ where (x_1, y_1) and (x_2, y_2) represents the points. Print the distance between the points (14,16) and (38,63).

84. Use the distance formula to determine the distance between the points (15,16) and (30,48).

85. Find the distance between any two given points (a,b) and (c,d). Use the following points as data: (2,3) and (4,7); (1,8) and (–2,8); (–3,6) and 12,–4); (–7,0) and (0,2).

86. Input the coordinates of two points A and B in the coordinate plane. Determine the length of the segment AB.

87. Input the coordinates of two points A and B in the coordinate plane. Find the coordinates of the midpoint of the segment AB.

88. Use the coordinates of two points in the plane and compute the distance between them, the coordinates of the midpoint, and the slope of the line segment.

89. Given the coordinates of four points in the xy-plane, decide whether the quadrilateral formed by joining the points in order is a parallelogram.

90. Give the coordinates of three points in the XY-plane, determine whether the points are collinear.

91. Determine the circumference of a circle with any given diameter. Test your program for the following diameters: 4, 100, 16.4, 34000.

92. The radius of the earth is about 7400 kilometers. Calculate the circumference of the earth.

93. What is the area of a circle with a radius of 8 centimeters?

94. Input the radius of a circle, R. Determine the area by using 3 1/7 for π and 3.14159 for π. Input several values of R and print the results of computation in a tabular format.

95. Given the coordinates of the center and the length of its radius, determine the equation of a circle.

96. What will happen to the area of a circle if the measure

of a radius is doubled? Halved? Tripled? Write and run a program that will help you answer.

97. A cylinder is 1.1 meters long and the radius of its base circle is 7 centimeters. What is its volume (a) in cubic centimeters, (b) in cubic meters?

98. A farmer plants his peanuts in a 61 meter diameter semicircular field. Write a program to determine the area of the field.

99. Given that a circle passes through (2.1,–.3), (.1,.5), and (1.02,–.03), find the coordinates of the center and the measure of the radius.

100. Find the area enclosed by the graph of any circle of the form $x^2 + y^2 = r^2$.

101. Tom ordered 36 meters of fencing for a rectangular dog pen. Many rectangles have 36 meter perimeters; for example 6 × 12, 8 × 10, and 9 × 9. Determine the rectangle that has the largest area for his dog.

102. A farmer owns a large piece of property that borders on a straight river. He has 30 meters of fencing, and he wishes to fence off a rectangular area, using the river as the boundary along one side of the rectangle and the fencing along the other three. Find the shape of the rectangle with maximum area. What is the shape with maximum area if fencing can be installed only in 3 meter lengths?

103. Determine the geometric mean of two positive real numbers.

104. The numbers P(N) = 4(1 – 1/3 + 1/5 – 1/7 + ... 1/(2N – 1) are known to converge to π. Compute P(N) up to P(1000), printing every hundredth value. The thousandth should turn out to be 3.140578, which is off in the third decimal place.

CHAPTER 4

TRIGONOMETRY

Contained in this chapter are problems suitable for use by students in a typical trigonometry course or in other mathematics courses which include trigonometry as an integral part of the course.

1. Convert measures from degrees to radians, using multiples of 10° from 0° to 360°.
2. Input the degree measure of an angle and compute the radian measure.
3. Input the radian measure of an angle and determine the degree measure.
4. Find the angles of a 3, 4, 5 right triangle to the nearest minute.
5. Find the angles of a 5, 12, 13 right triangle to the nearest minute.
6. Any angle whose degree measure is greater than 90 or less than 0 has a "reference angle" between 0 and 90, inclusive. Input the degree measure of an angle between -360 and 360, inclusive, and print the measure of its reference angle.
7. Convert from polar coordinates to rectangular coordinates the following polar equations:

   (a) r = cos 3g
   (b) = sin 3g
   (c) r = sin g + cos g

8. Find the value of sin x and cos x for x = 30°, 45°, 60°, 90°.

9. Without using preprogrammed library functions for the sine and cosine, generate a table for the sines and cosines of all angles between 0° and 90°.

10. Write a program which will output in column form, the sine, the cosine and the tangent of X, where X is in degree measure. Input your starting angle A, the increment I, and the final angle B.

11. If the sides of a triangle are 10, 10, and 4 meters, find the angles of the triangle to the nearest minute.

12. A right triangle has one angle 42° 25′ and the side opposite that angle has a length of 25.4 centimeters. Find the other sides of the triangle.

13. Determine the area of a triangle by taking one-half the product of two sides times the sine of the included angle.

14. Input the lengths of the hypotenuse and one leg of a right triangle. Determine the sine, cosine and tangent of either of the acute angles of the triangle.

15. Read the lengths of the legs of a right triangle. Compute and print the values of the six trigonometric functions of either acute angle of the triangle.

16. Print $\sin^2 x + \cos^2 x$ for x = 5°, 10°, 15°, 20°, ..., 85°. Examine the printed output. What conjecture can you make?

17. Write a program for checking the validity of the trigonometric equation $\sin 2\theta = 2 \sin \theta \cos \theta$ for ten values of θ.

18. If we know the lengths of two sides of a triangle and the measure of the included angle, we can also use the "Law of Cosines" to determine the length of the third side. For any triangle ABC, the Law of Cosines gives these three relationships:

$$a^2 = b^2 + c^2 - 2bc \cos A$$
$$b^2 = a^2 + c^2 - 2ac \cos B$$
$$c^2 = a^2 + b^2 - 2ab \cos C$$

Use the Law of Cosines to find the length of side a of a triangle with sides b = 6, c = 8, and an included angle of 22°.

19. The Law of Sines states that for any triangle ABC, a/Sine A = b/Sine B = c/Sine C. Given the measures of the sides a and b and angle C, and using the Law of Cosines, determine the measure of side c and angles A and B. The Law of Cosines states that for any triangle ABC, $c^2 = a^2 + b^2 - 2ab$ Cosine C.

20. Two speedboats leave a boat dock at the same time. One boat travels due north at 57 kilometers per hour, and the other travels 40 degrees west of north at 63 kilometers per hour. How far apart are the two boats after 2 hours?

21. A senior at Satellite High School wants to know the height of the flag pole in front of the office building. His measurements show that a distance of 183 meters from the base of the pole, and the elevation angle of the top end is 22.5°. Write a program to determine the height of the pole.

22. Compute the area of a regular polygon of N sides, each of length L.

$$\text{Area} = \tfrac{1}{2} N^2 L \cot(180°/N)$$

The program should input values for N and L.

23. Find the resultant force acting on P, given the following information: Angle CWB = 41°, Angle CWA = 72°.

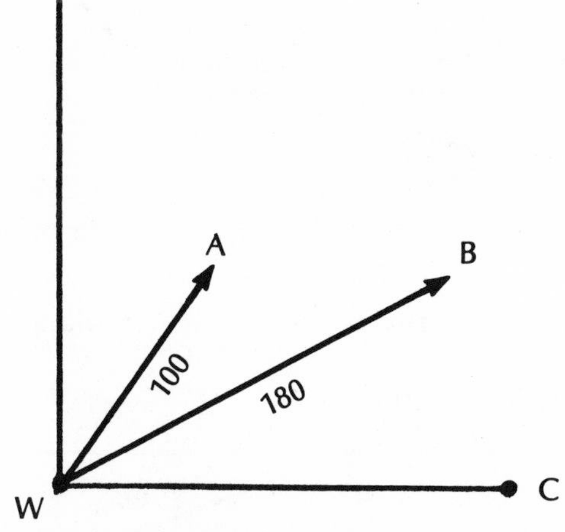

24. Standing 274 meters from the base of a lighthouse on level ground, the angle of elevation is 8° 15′. Find the height of the lighthouse.

25. Rusbox and Ashville, a civil engineering firm, is building a bridge across the Tomoka River from point A to point B. To find the length of the bridge, an engineer puts stakes in the river bank at points A and B. He then locates a point C, 30 meters from A, such that triangle BAC is a right triangle. He determines that angle BCA is 55°. How long must the bridge be to span the river?

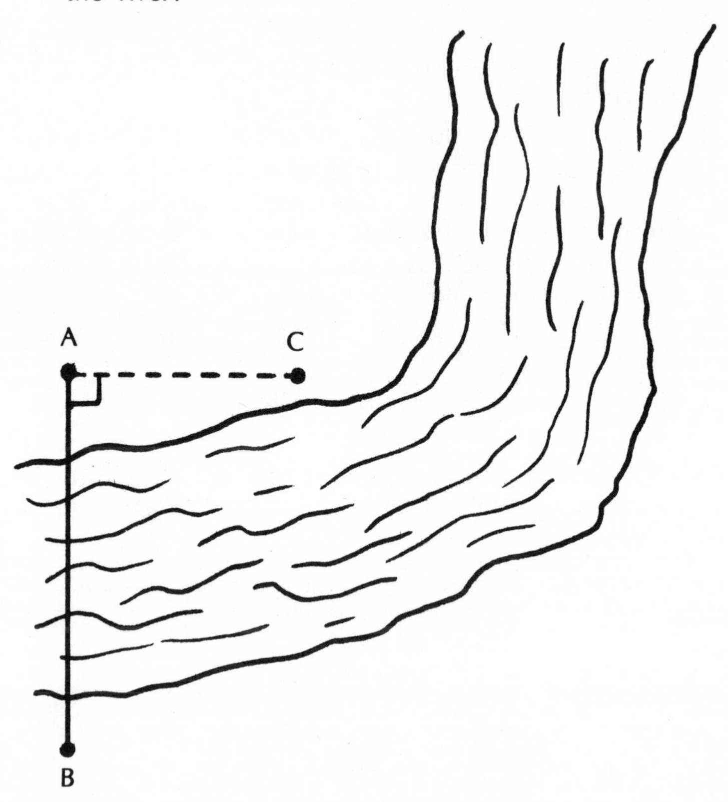

26. Find the area of the following triangle.

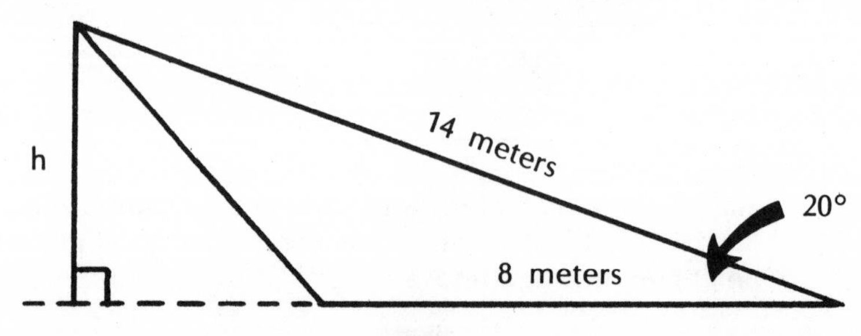

27. A real estate salesman wanted to measure the length of the lake shown. In order to find the length AB, he located a point C, 95 meters from A and 122 meters from B. He found the measure of angle ACB to be 47.5°. How long is the lake?

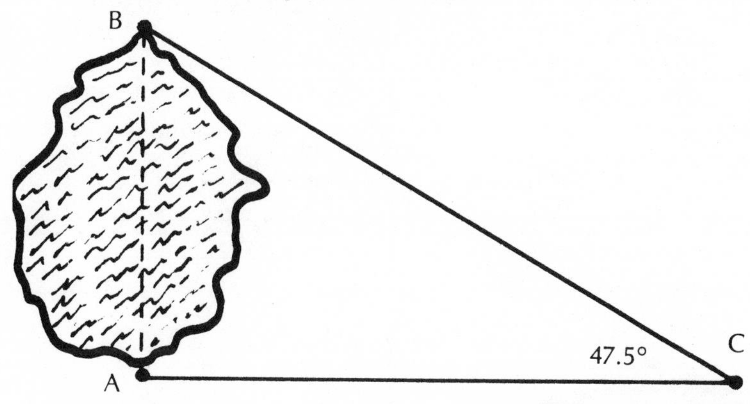

28. At Twin Rivers Park, there is a walking bridge from X to Y. The park directors want to add bridges from X to Z and from Y to Z. How long would these two bridges be?

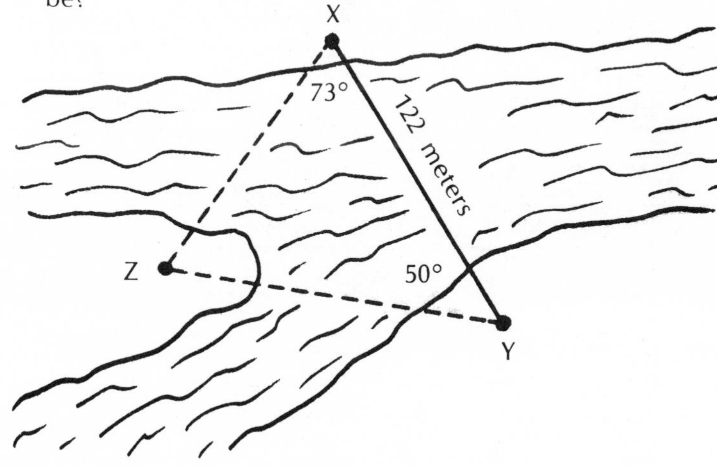

29. Compute sin x using this series: $\sin x = x - x^3/3! + x^5/5! - x^7/7! + \ldots$ where x is in radians.

30. Compute cos x using this series: $\cos x = 1 - x^2/2! + x^4/4! - x^6/6! + \ldots$ where x is in radians.

31. Compute arctan x using the Taylor series: $\arctan x = x - 1x^3/3 + 1x^5/5 - 1x^7/7 + \ldots$ where $-1 < x < 1$.

32. A trigonometric function states that for all values of x: $\sin^2(x) + \cos^2(x) = 1$. Write a program designed to check this equation. (Hint: In your program, set $y = \sin^2(x) + \cos^2(x)$).

33. Compute the area of a segment of a circle by the formula $\text{Area} = \pi r^2/2 - [x\sqrt{r^2 - x^2} + r^2\sin^{-1}(x/r)]$, where r is the radius of the circle and x is the perpendicular distance of the chord from the center.

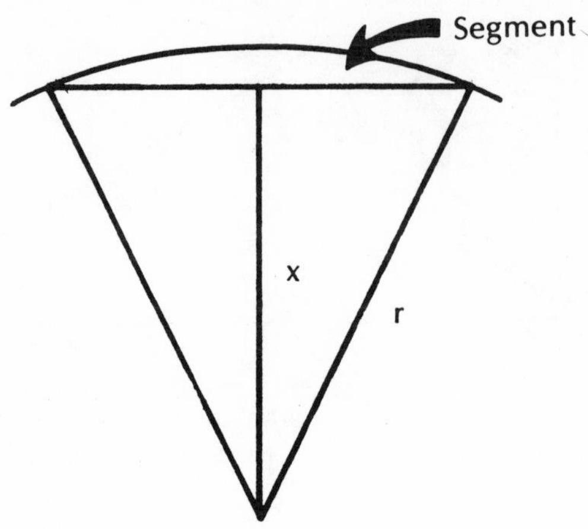

34. A Pythagorean triple is a set of positive integers A, B, and C such that $A^2 + B^2 = C^2$. Determine all Pythagorean triples whose components are all less than or equal to 50.

CHAPTER 5

PROBABILITY AND STATISTICS

In recent years, high schools and colleges have reported an increased interest in courses in probability and statistics. It has become increasingly apparent that knowledge in this area is indispensable for the effective pursuit of studies in psychology, sociology, business administration, economics, game theory, medicine, political science, geology, and other disciplines.

Central to a course in Probability and Statistics are many exercises involving lengthy calculations. A computer can help students perform many of these tedious and time consuming calculations.

The problems found in this chapter may be used by students in any course in Probability or Statistics, or in any course where probability and statistical concepts are taught.

1. Simulate the tossing of a coin.
2. It is a well-known fact that each time a coin is flipped, there is a 50-50 chance that it will come "tails". Write a program to print a typical sequence of 100 flips (that is: TAILS, HEADS, HEADS, TAILS, etc.).
3. Have the computer simulate the flipping of six coins 1000 times and print the distribution of outcomes.

4. Simulate tossing a coin 5 times in succession. Repeat the sequence of 5 tosses 100 times counting the number of heads appearing in each sequence of 5 tosses. When this is completed output the number of times no heads appeared, 1 head appeared, 2 heads appeared, 3 heads appeared, 4 heads appeared, and 5 heads appeared.
5. Write a program which will simulate tossing a regular six-sided die 60 times. Count and print the number of times each side comes up.
6. Simulate 1000 rolls of a die. Do not print the result of each roll. At the end of the simulation print the proportion of fours which turned up.
7. Roll a die 1000 times. Count the number of times the 3 comes up.
8. Write a program to simulate the rolling of two dice 1000 times and output the number of 7's and the number of 11's. Are the results reasonably close to what you expected?
9. Determine the percentage of times the sum of two dice will be 2, 3, or 12.
10. Four dice are rolled. Analyze the four numbers which appear on the top faces of the dice and determine whether 0, 2, 3, or 4 of the dice show the same value, or whether two show one value and the other pair show another value.
11. Five dice are rolled. Any die that shows a 4 is set aside and the others are rerolled. Any new 4's obtained on the second roll are also set aside and the remaining non-4's are rolled again. This process is continued until

all the dice are showing 4's. Approximately how many rolls are needed on the average?

12. Simulate the dealing of five-card hands from a standard 52-card deck. Be sure not to deal the same card twice.

13. If two cards are drawn from a standard 52-card deck of playing cards, what percentage of the time will the cards be an ace and a face card (king, queen, or jack)?

14. A roulette wheel is marked with even numbered slots from 2 to 36, odd numbered slots from 1 to 35, and the slots 0 and 00. Simulate 1000 spins of the roulette wheel and determine the proportion of times an odd number is spun.

15. A set of numbers which appears often and significantly in probability and analysis is N factorial (N!).

$$N! = 1 \times 2 \times 3 \times 4 \times 5 \times \ldots \times (N - 1) \times N$$

Generate a table of factorials up to some specified value of N.

16. Generate 100 days of weather in Wonderland by simulation. For example, if today is rainy, generate a random number X; if X = 1, then it rains again; if X = 2, then it is nice; otherwise, it will snow. Compute the days of each kind of weather and compare the results with the limiting probabilities.

17. Determine the probability of a number between 2 and 100 being a prime number.

18. A science class of 31 has 18 girls. A committee of five is selected at random. What is the probability that all five committee members are girls?

19. What is the probability that a non-leap year will have 53 Fridays?

20. A company makes bolts. It is known that 1 in 1000 is defective. You buy a box of 100 bolts. What is the probability of getting exactly one defective bolt?

21. Two dice are rolled until a 3 or 7 appears. What percentage of the time will the 3 be rolled before a 7 is rolled?

22. Pick two numbers at random between 1 and 20. What is the probability that the sum is 12?

23. Simulate the rolling of 3 dice and determine the probability that at least one of the three dice shows a three.

24. Select the suit of diamonds from a deck of 52 cards. What is the probability of being dealt the 4, 5, 6, 7, 8 of diamonds in that sequence?

25. Find the probability that more than 70% of the results will be tails if a fair coin is tossed 200 times.

26. Suppose the letters C, E, M, O, P, R, T, U are selected at random.What is the probability that the order of the letters will produce the word COMPUTER?

27. Cannibals A and B are both 50% marksman with the blowgun. They fight a duel where they exchange alternate shots. If Cannibal A shoots first, what is the probability that he will win?

28. A poker hand is dealt. Find the probability that the hand contains at least one pair, given that it contains no aces, tens, or face cards (king, queen, jack).

29. Medical records indicate that 40% of the cases of a particular ailment are fatal. If the Medical Center admits 8 patients suffering from this ailment, what is the probability that at least two will be cured?

30. A die has the number 1 on two opposite faces, the number 2 on two opposite faces, and the number 3 on two opposite faces. The die is rolled 500 times. Find the probability that the sum of the outcomes is greater than 1020.

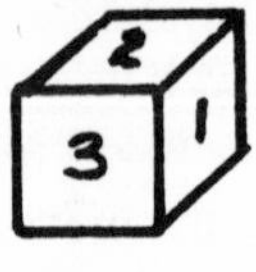

31. A gambler enters a Las Vegas casino with $1,000 and bets $1 on black at the roulette table each minute. What is the probability that he will have $1,000 or more after 1 hour?

32. When an American roulette wheel (0 and 00) is spun, what is the probability that the result is (a) 0, (b) 00, (c) 0 or 00, (d) even, (e) in the first 12, (f) in the second column, (g) 4, 5, 6, 7, 8, or 9?

33. Willie Bigstep, the catcher for the Houston Red Socks, has a probability of .3 of getting a hit each time he comes to bat. What is the probability in 50 times at bat that his average will be less than .250?

34. A politician running for office drops 10,000 leaflets on a city which has 2000 blocks. Assume that each leaflet has an equal chance to land on each block. What is the probability that a particular block will receive no leaflets?

35. Mathematicians claim that in a random group of about 23 people, the chances are about 1 in 2 that two or more of the people will have the same birthday (month and day). Write a program to select 23 birthdays at random and have the program determine whether any two of them are on the same date. Run the program several times.

36. You are in a room with 29 other people. What is the probability that one of them has your birthdate?

37. Compute and print a table of the theoretical probabilities that 2 people in a room full of N people have the same birthday. Vary N from 1 to 50.

38. Twelve people are in a room. Use a computer simulation to determine approximately the probability that at least three of them have birthdays in the same month.

39. Suppose we spin a spinner that is equally divided into three regions numbered 1, 2, 3. In this experiment there are only three possible outcomes, 1, 2, or 3. The theoretical probability of obtaining any one of these outcomes is 1/3. Simulate the spinning of a spinner 1000 times. Is the computed result close to the theoretical probability?

40. The king's minter boxes his coins 500 to a box, and puts one counterfeit coin in each box. The king is suspicious, but instead of testing all the coins in one box, he tests one coin chosen at random out of each of 500 boxes. What is the probability that he finds at least one fake?

41. A poker hand is a set of five cards randomly chosen from a deck of 52. Find the probability of (a) a royal flush, (b) a straight flush, (c) four of a kind, (d) a full house, (e) a flush, (f) a straight, (g) three of a kind, and (h) two of a kind.

42. Suppose a gambler bets $5 on the following game. Two dice are rolled. If the result is odd, the bettor loses. If the dice are even, a card is drawn from a standard deck. If the card is 1 (an ace), 3, 5, 7, or 9, the bettor wins the value of the card, otherwise he loses. What, on the average, will the bettor win (or lose) on this game?

43. Each January 1 in a certain western city the air pollution index is 100. Smoggy days and clear days occur "randomly", the probability of each being 1/2. On a clear day the index decreases by 10 percent of its value. Simulate 15 years of 365 days each to estimate the probability that the pollution index is greater than 105 on any given day.

44. If two unbiased coins are tossed, the probability of obtaining two heads is 1/4. If this experiment is repeated 10 times, the probability of obtaining two heads exactly K times is

$$\frac{10!}{k!(10-k)!} (1/4)^k (3/4)^{10-k}.$$

Compute a table of values of this probability for k = 0, 1, ..., 10 and determine which value of k is most likely.

45. Bill had five girl friends, and one evening he wrote a letter to each of them. He also addressed five envelopes. Bill's young sister is always trying to do nice things for Bill, so she put the letters in the envelopes and mailed them. If she chose the envelope for each letter at random, what are the chances that not one of the letters was put into the correct envelope?

46. Two witches enjoy meeting each evening over a cauldron of tea, but both witches have two serious shortcomings. First, each witch is poorly organized and arrives at the meeting place randomly between midnight and 1 a.m. Second, each is notoriously evil-tempered and becomes enraged upon having to wait 15 minutes or longer for her companion. Thus, the following temper-saving arrangement has been agreed upon: When either witch has waited 15 minutes — or when 1 o'clock arrives and she is still alone — she disappears at once, not returning until the next night. Here is the problem: On a given night, what is the probability that the two witches meet?

47. In some games numbered or colored objects are drawn from a container in which the original collection of these objects is first shaken very thoroughly. Suppose there are five white and ten black balls in a bag and three of these balls are to be drawn without replacement. Simulate these draws on a computer and use it to determine approximately the probability of drawing exactly one white ball and two black balls.

48. A secretary has two cars, but neither is in very good condition. At any given time, the probability that a car will start is 7/10 for either car, considered independently. Happily, once a car has been started, it will continue in operation all the way to work or back home, as the case may be. In addition, the secretary has a boyfriend who will sometimes drop by to take her to work or to take her home. If he arrives, she does not even try to start her car(s), but the probability that he will come by is only 1/2. It is quite possible that the secretary will find herself at home in the morning without any transportation, because the boyfriend did not come by, she could not get a car started, or both cars were at work. In this case, she calls her boss. He is not sympathetic, and fires her, whereupon she calls her boyfriend for consolation. He marries her, and they honeymoon in Paris. Analogously, it is possible that she will be stranded at work in the evening. In this case, she will elope to Mexico with the bookkeeper. Given that she starts at home with both cars, what is

the expected duration of her employment, and what is the probability that she will get to Paris?

49. For the month of August the local weather bureau has figures to show that on any given day the probability of rain is .20 if it has not rained on the day before, and .60 if it has. Simulate the rainfall pattern for a given ten day period in August for 100 years. Assume that on the first day of the period the probability of rain is .20, that is, that there was no rain the day before. Calculate the probability that there will be no more than three days of rain in this ten day period.

50. A gin hand consists of 10 cards from a deck of 52 cards. Determine the probability that a gin hand (a) has all 10 cards of the same suit, (b) has 4 cards in one suit, 3 in another, 2 in another, and 1 in a fourth suit, and (c) includes all 4 cards of each of the two different face values.

51. Write a program to calculate the probability that, on a given day in a given maternity ward, ten babies born will all be boys. Use the binomial distribution.

52. Mr. X and Mr. Y are about to fight a duel in which each takes turns shooting at the other. X hits whatever he shoots, on the average, once in every two tries. Y hits once in every three tries, on the average. Being a gentleman, X naturally allows Y to shoot first. Simulate the fighting of this duel and try to determine who has the better chance of surviving.

53. Simulate the playing of a set of tennis. Set the probability of player A winning a point equal to P, where P is given as data. Run 100 sets each for varying P's, and see how the probability of winning the set depends on the probability of winning a point.

54. Simulate the rise-or-fall performance of the stock market average over a period of 10 market days. The market rises with a probability of 0.30 if it fell the previous day, and with probability of 0.60 if it rose the previous day. Assume that the market rose the day prior to the start of your simulation.

55. In a class there are 20 students, 6 are age 18, 10 are age 19, and 4 are age 20. A student is chosen at random. What is his expected age?

56. A card is drawn at random from a deck of cards. If it is a red card a player wins one dollar. If it is a black card he loses two dollars. What is the expected value of the game?

57. A lawyer from Phoenix never puts money in a five-cent parking meter. He assumes that there is a probability of .05 that he will be caught. The first offense costs nothing, the second costs 50 cents, and subsequent offenses cost one dollar each. Under his assumptions, how does the expected cost of parking 20 times compare with the cost of putting money in the meter each time?

58. Print all permutations of N things taken N at a time for all N less than or equal to 10.

59. Compute the number of permutations of N things taken R at a time.

60. Compute the number of combinations of N things taken R at a time.

61. List all possible arrangments of the first 3 integers.

62. In how many ways can 15 people enter a classroom?

63. Write a program that will print all combinations of 4, 5, and 6 letters taken from a six letter word. Do not repeat any letters in the same word. Try your program on the word "NUMBER".

64. How many possible batting orders are there for a baseball team of 9 players?

65. Boats come in 12 colors, 5 models, 7 engines, and there are 8 options such as canvas top, ski rack, fish box, ship-to-shore radio, etc. How many different boats are available?

66. How many words can be formed using all the letters in COMPUTER?

67. A student has four algebra books, three English books and two American history books. She wishes to keep all books of the same kind together on her bookshelf. How many ways can the books be arranged?

68. You have five different flags with which to form signals by arranging them all on a flagpole. How many signals can you form?

69. How many different five-card hands may be dealt from a standard deck of 52 cards?

70. In how many different ways can 12 keys be put on a circular key ring?

71. You have 27 different books and two bookshelves, one of which holds exactly 13 books and the other holds exactly 14 books. In how many ways can the books be arranged on the shelves?

72. A family of eight, Father, Mother, Sandra, Susan, Sherrie, Steven, Laura, and Michael, is trying a different seating arrangement at a table for eight people every day. Determine all possible seating arrangements.

73. Determine all possible seating arrangements of a family of six, Father, Mother, Tom, Bill, Nancy and Ruth, at a table for eight people.

74. A Boston Union Hall has just acquired five new members who can be trained for five available jobs. In how many different combinations can the trainees be placed in the various jobs?

75. Roll 8 dice 600 times. Count the number of 5's that come up for each roll. Print the distribution.

76. An experiment consists of flipping a coin until it comes up heads. Write a program to perform the experiment 2000 times and count the number of flips required for each. Print the distribution.

77. The arithmetic *mean* is the sum of all values divided by the number of values. The *median* is the "middle value". Half of the values are larger than the median and half of the values are smaller. The *mode* is that value which occurs most frequently. Determine the mean, median, and mode of the following set of values: 153, 158, 161, 157, 150, 153, 149, 153, 155, 162.

78. Find the average investment in the bank if the bankbook recorded the following values on the first day of each year: $1000, $1040, $1081.60, $1124.86.

79. Find the average of 1000 random numbers.

80. Your class in ecology has five members who make the following scores on the final exam: 75, 93, 41, 98, and

71. Your teacher wants to compute the average exam score. Write a program to do the computation.

81. Find the average of a list of nonzero numbers, and then print the following four quantities: the average, the number of these numbers which are greater than the average, the number equal to the average, and the number of them less than the average.

82. The following scores were recorded on the Mathematics College Entrance Exam: 83, 74, 69, 100, 92, 95, 89, 75, 92, 82, 85, 97, 74, 91, 78, 83, 61, 100, 93, 54, 87, 82, 79, 68, 72, 75, 86, 92, 53, 100, 99, 67, 97, 79, 82, 81, 85, 98, 99. Determine and print the median of the scores. Also print the scores in descending order.

83. The Wilson family went on vacation last week. They drove 440 kilometers on Monday, 0 kilometers on Tuesday, 100 kilometers on Wednesday, 320 kilometers on Thursday, and 40 kilometers on Friday. Determine the average distance traveled per day.

84. Find the standard deviation for a set of data. Use the following set of data in your program: 220, 180, 275, 200, 240, 215, 208, 197, 223, 189, and 218.

85. Arrange the numbers 93, 81, 97, 75, 69, 92 in descending order and compute their mean, sum of squared deviations, variance, and standard deviation.

86. Write a program which will find the mean and standard deviation of a given sample.

87. Compute the average for and variance of the gas mileage of a late model Chevrolet which traveled 200 miles, 180 miles, 155 miles, 230 miles, 143 miles, and 190 miles, respectively. The gallons of gas used for each trip are 13.2, 10.8, 9.6, 12.5, 9.5, and 11.6. The formula for computing the variance is

$$\text{Variance} = (\text{average} - X_i)^2/(N - 1)$$

where $i = 1$ to N, N is the number of trips and x_i is the miles traveled for trip i.

88. An automobile dealer's sales of new cars are distributed uniformly between zero and nine cars inclusive, daily. Simulate sales over a 15-day period.

89. Find the sum of 10,000 random integers 0, 1, 2, ..., 9. Compute the average of the numbers obtained. Have

the program tested to see whether the average lies within 3 standard deviations of the expected value of 4.5.

90. The college board scores of ten students in algebra and English are given below. Compute the mean of the algebra scores, the mean of the English scores, the standard deviation of the algebra scores, the standard deviation of the English scores, and the coefficient of correlation. Algebra scores - 750, 770, 740, 700, 680, 710, 700, 750, 720, 680. English scores - 620, 680, 600, 710, 700, 690, 700, 708, 675, 710.

CHAPTER 6

INTERMEDIATE MATHEMATICS

The problems presented in this chapter may be used in advanced mathematics classes in high school or in college mathematics courses. The reader will find a wide selection of problems from the disciplines of algebra, number theory, calculus, numerical analysis, theory of equations, matrices, trigonometry, geometry and linear programming.

1. Have the computer convert numbers in base 10 to base 8.
2. Convert a positive integer in base 10 to any base B.
3. Convert a positive integer from any base B to base 10.
4. Input a number, base 2, having 6 or fewer digits and convert it to an equivalent number, base 10.
5. Write a program to add and subtract any two given Roman numerals (less than MM). The results are to be printed in the following format:

   DCCCLX + CDIV = MCCLXIV
   XXVI - XIV = XII
6. Convert Roman numerals to Arabic.
7. Convert Arabic numerals to Roman numerals.
8. Given the coordinates of a point in the Cartesian plane, convert to polar coordinates.

9. Input the polar coordinates of a point and determine the equivalent point in Cartesian coordinates.
10. The equation 16/64 = 1/4 is a result which can be obtained by the cancellation of the 6 in the numerator and denominator. Find all the cases in which AB/BC = A/C for A, B, and C integers between 1 and 9 inclusive. Do not consider obvious special cases such as 22/22, 33/33, etc.
11. Calculate the sum of the square roots of the odd numbers between 1 and 1000.
12. Print all numbers up to 1100 which are not divisible by any integer less than 10.
13. Print the first 25 terms of the sequence 3, 5, 6, 25, 9, 125, ...
14. Given a year, determine the next year in which January 1 will fall on the same day of the week.
15. Input a four-digit integer which represents a year. Determine whether it is a leap year.
16. Determine how many times Friday the 13th occurs in a specified year.
17. Input a year, N, and print the calendar for that year.
18. Approximate the square roots of 7, 16, 48, 163, 1275, 78.5, and 401.32. The program should not use a preprogrammed square root library function.
19. Prepare a table of the values of e^x for x = 0, 0.01, 0.02, ..., 1.99, 2.
20. Find the maximum positive integer x for which the square root of x is less than 100 and to find the maximum positive integer x for which exp x is less than 100.
21. Given a list of 30 numbers, reorder or sort them into ascending order.
22. Using the following coins, half dollar, quarter, dime, nickel, and penny, how many different ways can you make change for a dollar?
23. Determine what value of N would make 14N5N divisible by 19.
24. Input three consecutive terms of a number sequence.

Determine whether the sequence is arithmetic, geometric, or neither.

25. Read the components of two two-dimensional vectors. Determine if the vectors are parallel and/or perpendicular.

26. Given pure imaginary numbers Ai and Bi, input A and B and compute the product of Ai and Bi.

27. Write a program to print the sum, difference, product, and quotient of two complex numbers a + bi and c + di. Use the following complex numbers as data:
    (a) 2 + 3i and 3 + 4i
    (b) 5 + 7i and 1 + 4i
    (c) −21 + 3i and 14 + 107i

28. Write a program that gives practice in adding and multiplying complex numbers.

29. Find the multiplicative inverse of the complex number 2 + 3i.

30. Find the quotient of two complex numbers in polar form.

31. Generate random complex numbers. Then test the associative property for both addition and multiplication.

32. If a population of three million doubles every five years, how many years will it take to reach 300 million?

33. Miss Murray, history teacher at Saint Mary's High School, uses the following grading scale on her tests: A = 100-93, B = 92-84, C = 83-74, D = 73-70, F = 69-0. First read N, which represents the number of students in a class. Then read the exam grade of each student and count the number of A's, B's, C's, D's and F's in the class.

34. For the next three weeks (Monday through Sunday) you are working at Joe's Hamburger Stand. Joe will pay you a penny on the first day, two cents on the second day, and four cents on the third day. Each day you are paid double the amount of the previous day. What will your salary be on the last day you work for Joe?

35. A pipe smoker has two booklets of matches in his pocket, each containing 40 matches initially. Whenever

a match is required he picks one of the booklets at random, removing one match. Simulate the situation 100 times (using random numbers) and determine the average number of matches that can be removed until one booklet is completely empty.

36. A rich man wants to give away a sum of money by dividing the sum equally among a number of needy families. (The sum is below $10,000, and he will calculate the equal division to pennies). He notes that if he kept a penny, he could divide it equally among 31 families; if he kept a nickel, he could divide it among 32 families; if he kept a dime, he could divide it among 33 families; and if he kept a quarter, he could divide it equally among 35 families. How much money does he have to give away?
37. A man who lives 40 blocks from his dentist's office gets a toothache. In the first 10 minutes he walks halfway from his home to the dental office, but because his memories of the drill become more and more vivid as he gets closer and closer to his destination, he begins to slow down. In each 10-minute interval he covers half the remaining distance. Write a program designed to determine where he is after one hour. (Hint: set up a loop with D = 1/2 * D and pass the calculations through the loop six times).
38. An airplane flying at altitude A passes directly over point P. If its speed is S, compute its distance from point P at times T = 1, 2, 3, ..., 60 after the pass.
39. The squares of the natural logarithms are used in predicting the tides. Write a program designed to compute the squares of ten of the natural logs.
40. A track coach has just received a set of hop, skip, and jump distances of the contestants in a track meet. In this event, the winner is the athlete who has the greatest combined total distance. Write a program to determine the winner of the event in which there were sixteen competitors.
41. A piece of wire 30 centimeters long is to be cut into two pieces. One piece is to be bent into the shape of a square and the other into a circle. Where would the wire be cut so that the sum of the two areas is the maximum?

42. Worldwide Airlines has made its flight schedule available to you. The schedule contains the following information for each flight: departure time and arrival time at destination. Determine how many flights last one hour, two hours, three hours, etc.

43. A section of city street is 150 meters long and it is marked off into parking places of 600 centimeters each, so that a total of 25 cars can be parked. If the section were not marked off and cars were allowed to park at random, approximately how many could be parked? (Assume that 600 centimeters are required to park a car).

44. The middle term of three successive terms in a harmonic progression is called the "harmonic mean" between two numbers. Determine the harmonic mean H between two numbers M and N using the formula $H = (2 \times M \times N)/(M + N)$.

45. A man can row at the rate of 5 kilometers an hour and walk at the rate of seven. If he is 100 meters offshore and wishes to reach a point that is 300 meters inland and 500 meters down the shore, what is the least amount of time he requires to reach his destination?

46. Arrange in ascending order any number of 3 digit numbers (up to 999). You may assume that no two numbers are equal.

47. This game tests your skills in binary-to-decimal and decimal-to-binary conversion. You will be given twenty conversion trials. Numbers are chosen randomly and your score will be printed at the end. The answer to any conversion you miss will be displayed; if the next conversion is presented, you may assume you got the previous one correct. Write a program that will play this game.

48. Suppose you were to fold a piece of paper several Times. Each successive fold should produce a piece of paper twice as thick as the previous one. Assuming that the thickness of an unfolded sheet of paper is .03 inches, determine the thickness of the paper after 35 folds. The program should convert inches to feet whenever 12 inches is exceeded and feet to miles whenever 5280 is exceeded.

49. An auto parts dealer keeps supplies of a certain item in his warehouse. He has found that the cost of doing so is .01x + 10/x dollars per month, where x is the number of items he orders every time his supply is low. Part of the cost increases as x increases because he has to provide more space to look after the larger deliveries he receives. On the other hand, part of his cost decreases as x increases because it is less expensive for him to place fewer orders. What is the most economical order for him to place?

50. Find the sum and last term of any arithmetic progression. Use the formulas:

$$l = a + d(n - 1)$$
$$s = 1/2n(a + l)$$

where a is the first term, d represents the common difference, n represents the number of terms, l represents the last term, and s represents the sum of the terms. Example: in the progression 1, 3, 5, 7, 9, – l = 9 and s = 25.

51. Input A, the first term, D, the common difference, and N, a positive integer of an arithmetic progression. Determine the Nth term of the sequence.

52. Find the sum and last term of any geometric progression. Use the formulas:

$$l = ar^{(n - 1)}$$
$$s = a - rl/(l - r)$$

where l represents the last term, a represents the first term, n represents the number of terms, r represents the common ratio, and s represents the sum of the terms. Example: in the progression 2, 6, 18, 54 – l = 54, and s = 80.

53. Input A, the first term, R, the common ratio, and N, a positive integer for a geometric progression. Determine the Nth term of the sum of the first N terms of the sequence.

54. Find the sum of the series n(n + 1)/2 for n = 1, 2, 3, ..., 20.

55. A sequence of numbers is defined as follows: The first two numbers are 1, 2. From any two consecutive numbers a, b, the next one is obtained as .5(a + b).

The sequence begins with 1, 2, 1.5, 1.75, 1.625, ... Print fifteen numbers in this sequence.

56. Print the first 30 terms of the sequence 1, 1/2, 1, 1/4, 1, 1/8, ...

57. Print the sum $1 + 1/2^2 + 1/3^2 + 1/4^2 + ... + 1/N^2$ for each N = 2, 3, ..., 1000.

58. A sequence of numbers is defined as follows: the first three numbers are 1, 1, 1. From any three consecutive members a, b, c the next is obtained from c + 2b + 3a The sequence starts with 1, 1, 1, 6, 11, 26, 66, ... Print twenty numbers in this sequence.

59. In a certain geometric sequence, the first term is 6 and the common ratio is 1/2. Print the first 20 pairs in the sequence.

60. Generate and print 200 random numbers between 1 and 12 and determine how often each number has occurred.

61. Write a program to generate 1000 random integers between 0 and 100. Determine how many of these numbers are "odd" and how many are "even".

62. Take the digits 1 through 9, written in increasing order, and insert between them either of the following three symbols: + (addition), – (subtraction), and blank. One example follows: 1 + 23 – 4 + 56 + 7 + 8 + 9 = 100. Determine all of the possible arrangements that produce a result of 100.

63. For each of the following pairs of numbers, find two numbers so that the sum of your two numbers is the first number in the given pair and the product is the second number in the given pair: 7, 12; 9, 18; 17, 60; –2, –8; –10, 21; 59, 168; 147, 572; 219, 1484.

64. After he has been drinking, Lucas, the town drunk, has an unusual way of walking. His steps are always exactly 1 meter in length, and he moves only in one of four directions, north, south, east, or west. After taking one step, the direction of his next step is chosen completely at random, but always in one of the four directions. There is a lamp post located on a pier 7 meters from the water on each of three sides and 9 meters from shore. If he starts from the lamp post,

what are his chances of returning to shore without falling into the water? Write a program to simulate Lucas' random walk. Let the walk end either when he has fallen into the water or made it to shore. In 50 walks, find the number of successful walks and the number of unsuccessful walks.

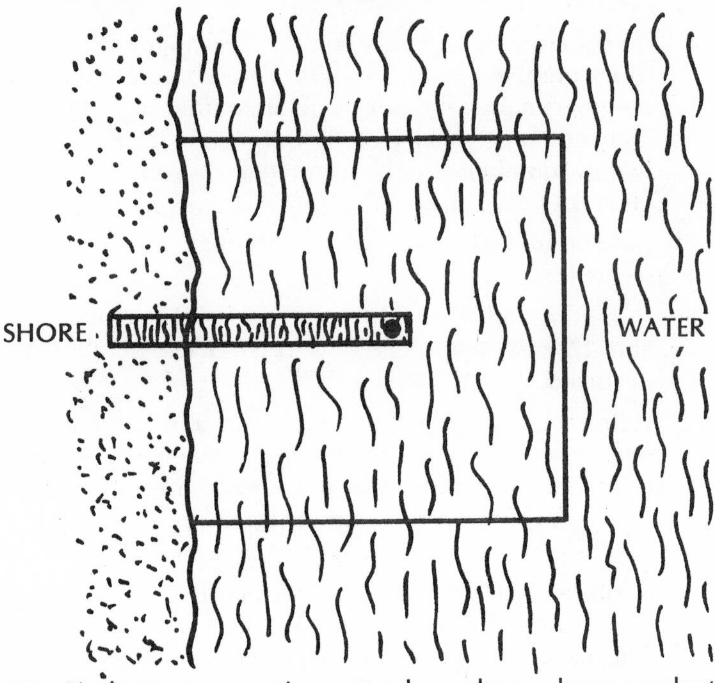

65. Find two consecutive natural numbers whose product is equal to the product of three consecutive natural numbers.

66. The irrational number π is one of the most interesting of all numbers.

$$\pi = 3.141592653589793238...$$

Surprisingly, several sums approach this value when the number of terms increase. For example, $\pi = 4(1 - 1/3 + 1/5 - 1/7 + 1/9...)$. Obtain an approximation of π using this series.

67. In 1773, a French naturalist named Buffon proposed an experiment for finding an approximation to π. He threw a needle of length L onto a large flat piece of paper ruled with parallel lines that were 2L units apart. He was able to show that the probability of the needle

landing across one of the lines is $1/\pi$. Produce an approximation to π by counting the number of crossings in a large number of trials.

68. Determine an approximate value of π using random numbers.

69. Determine which value is larger e^{π} or π^{e} . The value of π is 3.14159 and the value of e is 2.71828.

70. In a right triangle, the square of the hypotenuse equals the sum of the squares of the other two sides. Thus, in a 3, 4, 5 triangle, $5^2 = 3^2 + 4^2$. Find and print the sides of all such triangles if each of the 2 legs is integral, less less than 100, and each hypotenuse is a prime number.

71. Generate and print Pascal's triangle.

72. $70 is to be divided among 50 men, women, and children such that each man receives $6, each woman $3, and each child $1. Can this be done? If so, how many men, women, and children are there? (Hint: Use Diphantine equations in solving the problem).

73. A farmer can buy a turkey for $5, a duck for $7 and a chicken for $3. He wants to buy 30 birds for $100. How many of each kind can he buy? (Hint: Use Diphantine equations in solving this problem).

74. Compute the product AB (modulo M).

75. Have the computer print the addition table and the multiplication table for mod 5.

76. Have the computer do subtraction mod 8.

77. Construct a multiplication table modulo M. Print tables for M = 2, 3, 4, 5, ..., 10.

78. Print the first 20 rows of a table according to the following rules:
    1. The table is to have four columns labeled N, X, Y, Z.
    2. The values in the first row of the table are 1, 2, 3, 4.
    3. The values of N and Y are two greater than their values in the first row.
    4. The values of X and Z are four greater than their values in the first row.

79. Input 50 integers into an array, find the values of the

largest and smallest integers in the array, and print these two numbers and the fifty numbers in the array.

80. Interchange the two main diagonals of square number arrangement.

1	0	0	2
0	1	2	0
0	2	1	0
2	0	0	1

81. Our friends at the Western Computer Service Company have just lost their matrix multiplication capability, because of an unfortunate error in one of their system programs. The computer system is still working, but the matrix multiplication feature isn't. Write a program that will multiply a 4 × 5 matrix by a 5 × 6 matrix, assuming your own data.

82. Interchange columns 2 and 4 of the following number arrangement:

3	6	9	2	5
4	3	9	6	4
5	4	8	4	9
0	1	2	1	8
4	0	3	0	2

83. Determine if the following matrix has an inverse. If it does, the program should print the matrix with its inverse. If it doesn't, then have the computer print the message INVERSE DOES NOT EXIST.

$$\begin{vmatrix} 3 & -5 \\ -1 & 2 \end{vmatrix}$$

84. Find a root of the function $y = \sin x - x + x^3/6$ in the interval from $x = 0$ to $x = 2$.

85. Find a root of the function $y = \sin x - \cos x$ in the interval from $x = 0$ to $x =$.

86. Compute the zeroes of a polynomial using Newton's method.

87. Charles Babbage, the grandfather of the modern day digital computer, designed a difference engine to evaluate second order polynomials of the form:

$y = x^2 + 4x + 2$. Assuming that $x = 13$, write a program to evaluate this equation and print out the resulting value of y.

88. Find all of the roots of a cubic equation $AX^3 + BX^2 + CX + D = 0$, where A, B, C, and D are given as data.

89. Find the length of the curved line segment which belongs to the graph of $y = x^2$ between $x = 0$ and and $x = 1$.

90. The Taylor series $\cos(x) = 1 - x^2/2! + x^4/4! - x^6/6! \pm \ldots$ converges for all x. Compute the first 10 terms of this series.

91. Write a program to practice adding and subtracting polynomials.

92. Consider the system of equations,

$$Ax + By + C = 0$$
$$Dx + Ey + F = 0$$

Determine if the lines have the same slope, and if they do, whether they are parallel or the same line. If the lines are independent, determine their point of intersection.

93. Evaluate the 6th degree polynomial $y = x^6 - 3x^5 - 93x^4 + 87x^3 + 1596x^2 - 1380x - 2800$ for all integer values of x between −12 and +16. The program should print the heading VALUES OF X WHEN THE POLYNOMIAL IS ZERO and the values of x when y is zero.

94. Find the coordinates of points on $x + 4y = 4$ when x has the values 2, 2.5, 3, 3.5, 4, and evaluate $3x + 4y$ at each of these points.

95. Find a local maximum of the function sin x in the interval from $x = 0$ to $x = 2$.

96. Draw the plot of the curve defined by the equation $x^5 + y^5 + 5xy = 0$. The plot should include values of x from −2 to +2 in steps of 0.01.

97. Write a program to approximate the area of the region bounded by the x axis, the line = 0, the line $x = 3$, and the graph of the function: $f(x) = 3x^2/(1 + x^5)$. Do this by dividing the interval [0,3] into N sub-intervals, constructing rectangles whose bases are the sub-intervals and whose heights are the functional values

of the mid-points of the sub-intervals. Then sum these rectangles. Read in the values for N, and do the approximation for N = 6, 12, 15, 24, 30, 60, and 300.

98. Find the area under the curve $y = x^2$ between $x = 0$ and $x = 1$.

99. Compute the area under curve $y = x^3$ from $x = 2$ to $x = 10$.

100. Calculate the area under the curve $y = x^3$ between the limits $x = 1$ and $x = 4$.

101. Compute the area under the curve $y = 2x + 4$ from $x = 1$ to $x = 5$.

102. Find the area between the curve representing the function $y = (x + 1)/(x + 2)$ and the x-axis from $x = 1$ to $x = 2$.

103. Find the area between the curve representing the function $y = 1/x$ and the x-axis from $x = 1$ to $x = 10$.

104. The Taylor's series expansion of sin x is $\sin x = x - x^3/3! + x^5/5! - x^7/7! + x^9/9! - \ldots$ The error from truncation due to stopping at the Kth term is equal to or less than the absolute value of the Kth term. Write a program to calculate the sine of a given argument. The program should input the allowable error. Compare your result with standard tables.

105. If N is a large positive integer, an approximate value of N factorial is given by Stirling's approximation, $N! = \sqrt{2\pi N} \times N^N \times e^{-N}$. Using this formula, compute a value for 13!. Compare your computed value with the real value.

106. Find the area under the curve of a function by Simpson's Rule.

107. Write a program which does integration by the trapezoidal rule.

108. Find the points of intersection of the circle $(x - h)^2 + (y - k)^2 = r^2$ and the line segment that joins the points (a,b) and (c,d).

109. Find the points of intersection of the circle $x^2 + y^2 = r^2$ and the conic section with equation $ax^2 + by^2 = c$. The program should input values for r, a, b, and c.

110. Solve any pair of simultaneous equations in two variables.

$$ax + by = c$$
$$dx + ey = f$$

111. Compute the distance between points A and B in a three-dimensional coordinate system.

112. Given the following set of linear equations

$$3x + 6y + 7z = 6$$
$$4x + 27 - 3z = 7$$
$$6y - 2z = 2$$

Input the coefficient matrix and calculate and print its transpose and inverse.

113. The square root of a positive number can be calculated using the Newton-Raphson recursive method: $X_n = \frac{1}{2}(X_o + A/X_o)$ where A is the number and X_o is the first guess. Input values for A and X and compute the square root of A.

114. Compute the solutions of a cubic polynomial of the form $ax^3 + bx^2 + cx + d = 0$. (You may wish to reference a book on the Theory of Equations before attempting this problem).

115. It is often useful to find the equation of a curve that best fits a given set of points. One way of finding such a curve follows. In the figure, P and Q are two of a given set of points and $y = f(x)$ is a possible curve of best fit. P′ and Q′ are points on the curve having the same x-coordinates as the points P and Q. f(x) is chosen so that the sum of the squares of all the lengths PP′ and QQ′ is a minimum.

(a) *Linear case.* The equations for fitting a straight line $y = ax + b$ to a given set of points (X,Y) are to be:

$$a \Sigma X^2 + b \Sigma X = \Sigma XY$$
$$a \Sigma X + bN = \Sigma Y$$

and where N is the number of points.

(b) *Quadratic case.* The corresponding equations for fitting a quadratic curve $y = ax^2 + bx + c$ to a set of points (X,Y) are:

$$a \Sigma X^4 + b \Sigma X^3 + c \Sigma X^2 = \Sigma X^2Y$$
$$a \Sigma X^3 + b \Sigma X^2 + c \Sigma X = \Sigma XY$$
$$a \Sigma X^2 + b \Sigma X + cn = \Sigma Y$$

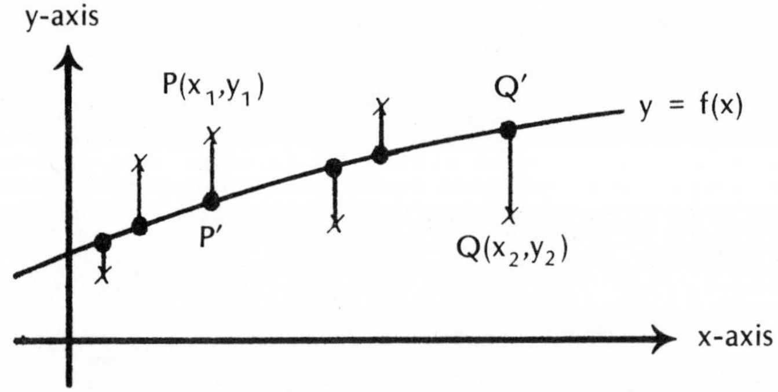

116. Write a program that finds the least common denominator (LCD) of three numbers.

117. A farmer has 20 chickens and 60 ducks. He wants to buy at least 100 more chickens and ducks, and he has space for at most 140 ducks on his farm. He wants to have at least as many ducks as chickens. He can buy the chickens at a cost of 15 cents each, and he can buy the ducks at a cost of 20 cents each. How many chickens and how many ducks should the farmer buy to minimize his cost?

118. Determine the minimum cost of supplying boxes from one of the warehouses to the store in Tampa.

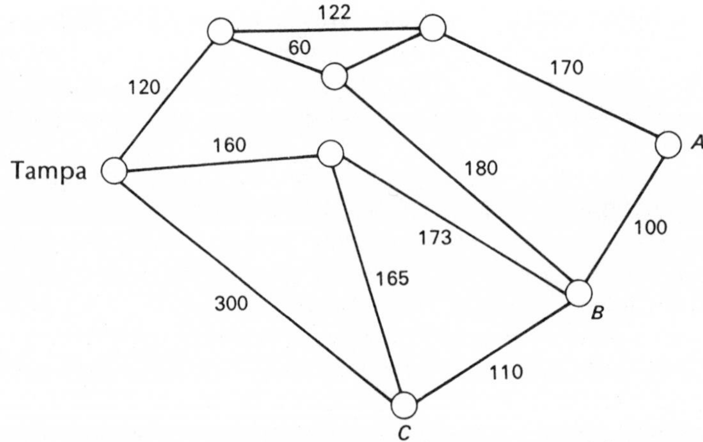

Shown is a map stating distances in kilometers between warehouses A, B, and C, and Tampa. Transportation costs and the cost of boxes are as follows:

	A	B	C
Box cost	.50	.47	.51
Transportation cost per box	.005	.006	.004

CHAPTER 7

NUMBER THEORY

Number theory is a branch of mathematics which deals with the *natural numbers*, 1, 2, 3, 4, 5, ..., often called the *positive integers*. Archeology and history teach us that man began early to count. He learned to add numbers and much later to multiply and subtract them. To divide numbers was necessary in order to share evenly a heap of pears or a catch of fish. These operations on numbers are called *calculations*. The word "calculation" is derived from the Latin *calculus*, meaning little stone; the Romans used pebbles to mark numbers on their computing boards.

As soon as men knew how to calculate a little, this became a playful pastime for many a speculative mind. Experiences with numbers accumulated over the centuries with compound interest, so to speak, till we now have an imposing structure in modern mathematics known as "number theory". Some parts of it still consist of simple play with numbers, but other parts belong to the most difficult and intricate chapters of mathematics.

If you are a puzzle enthusiast, you can probably recall many puzzles which depend for their interest upon properties of numbers, and if you have solved any of these puzzles, you have already been initiated into some of the elements of number thoery. Of all the unsolved problems in modern mathematics, a large proportion of them are to be found in the field of number theory. The problems found in this chapter are intended to introduce the student to this fascinating realm and to provide a framework for some nontrivial problems.

1. A "prime number" is any number that cannot be evenly divided by any other number except itself or 1. For example, 1, 2, 3, 5, 7, 11, 13, 17 are primes. Find all the primes between 2 and 400.
2. Input a number and determine if it is a prime number.
3. Input a number, N, and then list all prime numbers less than or equal to that number.
4. Print the prime numbers from 1000 to 1500. Your program should not test the even numbers.
5. The year 1973 will be remembered by many math buffs because it is a prime number. Generate all the prime numbers in the period 1972 — 2000.
6. Compute and print all the consecutive prime numbers (primes such as 11 and 13 or 21 and 23) less than 1000.
7. Compute and print a table giving the number and percent of prime numbers in the intervals 2—100, 101—200, 201—300, 301—400, ..., 901—1000.
8. List all composite numbers between 4 and 100. A composite number is any integer other than 1 which is not a prime.
9. Input two numbers, A and B, and determine if they are relatively prime. Two numbers are relatively prime if their greatest common divisor is 1. For example, 4 and 15 are relatively prime.
10. Input the three integers A, B, C and determine if they are relatively prime.
11. Determine whether a set of N positive integers are relatively prime.
12. Use the formula $3N^2 - 3N + 23$ to produce a table of prime numbers. Vary N from 0 to 22.
13. Print a table of prime numbers. Use the formula $P = N^2 - N + 41$ for N = 1, 2, 3, ..., 40.
14. The formula $Y_N = N^2 - 79N + 1601$ may be used to generate prime numbers in a limited range. Determine for each value N = 1, 2, 3, ..., 100 whether Y_N is a prime number or not.
15. A "palindromic prime number" is a prime that is also a prime when its digits are reversed. 17, 31, 37, and 113

are such primes. Find all the palindromic primes less than 400.

16. Generate prime numbers from the polynomial $x^2 + x + 41$ when x varies from 0 through 40.
17. It is possible to find arithmetic progressions of nonconsecutive prime numbers. For example, an arithmetic progression of three nonconsecutive prime numbers is 11, 17, 23 (constant difference of 6). Examine all prime numbers less than 1000 and print any arithmetic progression of three or four primes.
18. Mersenne primes are of the form $2^p - 1$ where p is a prime. For every Mersenne prime there is a corresponding perfect number $2^{p-1}(2^p - 1)$. Find several Mersenne primes and the corresponding perfect numbers.
19. Fermat, probably the greatest French mathematician of the 17th century, proposed that all numbers of the form $2^{2n} + 1$, where n = 0, 1, 2, ... are primes. Show that this is true for n = 1, 2, 3, 4, but not for n = 5.
20. It has been conjectured that there exists a large number of prime numbers of the form $n^n + 1$. For example, 1 + 1 = 2 (a prime) and $2^2 + 1 = 5$ (a prime). Determine for each value n = 1, 2, ..., 7 whether $n^n + 1$ is a prime number or not.
21. If p is a prime and $p^2 + 2$ is a prime, find a case where $p^2 + 4$ is also a prime.
22. Some prime numbers may be expressed in the form n! + 1. In fact it has been conjectured that there exists infinitely many prime numbers of this form. Determine which prime numbers (less than 2000) may be expressed as n! + 1.
23. In the thirteenth century, Leonardo Fibonacci, a wealthy Italian merchant who was fascinated with numbers, discovered what is now known as the Fibonacci number series. Each number of the series is the sum of the two numbers immediately preceding it. The first few numbers of the series are

    1, 1, 2, 3, 5, 8, 13, 21, 34, 55, 89, 144, ...

    Find the first 30 numbers in this series.

24. Suppose that two rabbits, one female and one male, are brought together to breed. Suppose that a pair of rabbits gives birth to another pair each month, beginning two months after their own birth, and assume that every pair produced in this way consists of one male and one female. How many rabbits will there be at the end of 24 months?

25. Find the first term of the Fibonacci sequence that is greater than 10,000. Only this term should be printed by the program.

26. Print the first fifty Lucas numbers. The Lucas numbers are similar to the Fibonacci numbers but the sequence begins differently: 1, 3, 4, 7, 11, 18, 29, 47, 76, ...

27. A perfect number is an integer such that the sum of its proper divisors is equal to the number. For example, 6 is a perfect number, since 1 + 2 + 3 = 6. See how many perfect numbers you can find.

28. Determine if a given number is perfect, abundant, or deficient. A number is perfect when the sum of the divisors of that number equals the number in question. A number is abundant when the sum of the divisors exceeds the number. A number is deficient when the sum of the divisors is less than the number.

29. Some even number pairs are called "amicable numbers" if the sum of the factors of each are equal to the other number. For example, 220 and 284 are amicable numbers since the sum of the factors of 220 is 284 and the sum of the factors of 284 is 220. Factors of 220 are: 1, 2, 4, 5, 10, 11, 20, 22, 44, 55, 110. (Sum = 284). Factors of 284 are: 1, 2, 4, 71, 142. (Sum = 220). Find all amicable number pairs less than 10,000.

30. A N- digit number is an Armstrong number if the sum of the Nth power of the digits is equal to the original number. For example, 407 is an Armstrong number because it has three digits such that $4^3 + 0^3 + 7^3 = 407$. Determine all Armstrong numbers with three digits.

31. The ancient Greek mathematician Diophantus is considered one of the earliest geniuses in number theory. However, Diophantine solutions to algebraic equations involved only positive rational numbers, and

in fact, modern usage restricts solutions to Diophantine Equations to positive integers. An actual equation as found in the writing of Diophantus reads, *Find three numbers such that their sum is a square and the sum of any two of them is a square.* Determine three such numbers.

32. Generate and print the first 15 rows of the Pascal triangle. Print the results in the format shown below.

1			
1	1		
1	2	1	
1	3	3	1

33. Given the table:

1	1	1						
1	2	3	2	1				
1	3	6	7	6	3	1		
1	4	10	16	19	16	10	4	1

What is the general rule after the first row? Use this rule to compute the next 10 lines of the table.

34. A Latin square is an N by N array of N numbers such that each number appears in every row and every column. Print all Latin squares of order N for N = 2, 3, ..., 9.

1	2	3	4	5
2	3	4	5	1
3	4	5	1	2
4	5	1	2	3
5	1	2	3	4

35. There are three integers less than 10,000 that are equal to the sum of their digits raised to the 4th power. An example is the number 1634. $1634 = 1^4 + 6^4 + 3^4 + 4^4$. Find the other two integers.

36. Find the greatest common divisor (GCD) of three numbers. The GCD is the largest number which divides all three numbers. For example, the GCD of 3, 12, 30 is 3.

37. Use the Euclidean Algorithm to determine the greatest common divisor of two positive integers.

38. Determine the greatest common divisor of the following number sets: 2059 and 4189, 18103 and 5473, 53053 and 689.

39. Of the N^2 possible pairs of integers between 1 and N, count the number of pairs whose greatest common divisor is 1.

40. Find the least common multiple of three numbers.

41. Find the least common multiple of five numbers.

42. Determine the greatest common divisor and the least common multiple of the following sets of integers: 324, 610; 200, 316; 84, 1003.

43. Determine all the different pairs of factors of the following set of integers: 1009, −654, 453, −991, 771, 991.

44. Input the number N and list its prime factors. For example, the prime factors of 12 are 2, 2, and 3.

45. Determine the number less than 2000 that has the largest number of factors.

46. Determine the smallest integer that has exactly 32 factors.

47. Find the sum and number of factors of: 1134, 1135, 1136, ..., 1174.

48. Divide 1000 into two parts so that one part is a multiple of 19 and the other part is a multiple of 47.

49. Write a program to play a game in which each of two players is to type a 7-digit number. The player whose number has the largest prime factor wins.

50. Find the possible values of N (less than 100) where the sum $1 + 2 + 3 + 4 + ... + N = N^2$.

51. The sum of the first five integers equal 15: 1 + 2 + 3 + 4 + 5 = 15. Which integers, beginning with 1 + 2 + 3 + 4 + ... would you have to add to get a sum of 820?

52. Determine what value of X would make X30X8 divisible by 23.

53. Plato, an Athenian teacher and pupil of Socrates, once recommended that a city be divided into plots of land so that the number of plots has as many proper divisors as possible. He suggested 5040 because it has

59 proper divisors. However, computers have determined that there are two numbers less than 10,000 which have 63 proper divisors. One of the numbers is 9240. What is the other number?

54. Given the sequence 1, 2, 4, 5, 7, 8, 10, 11, 13, 14, ... where every third integer is missing, print the sum of the first hundred terms of this sequence.

55. The numbers 12 and 13 have the following characteristic: $12 \times 12 = 144$, $21 \times 12 = 441$; and $13 \times 13 = 169$, $31 \times 13 = 961$. Find a three digit number (if one exists) for which this same kind of relationship exists.

56. Here is a mathematical curiosity in base 10: 98765432 times 9 equals 888888888. Find similar curiosities in bases less than 10.

57. What integer multiplied by 52631578947368421 will give all 9's?

58. Find a five digit number which when multiplied by four has its digits reversed. In other words, find a number ABCDE such that $4 \times ABCDE = EDCBA$.

59. Numbers which can represent a triangular pattern of dots are called *triangular numbers.*

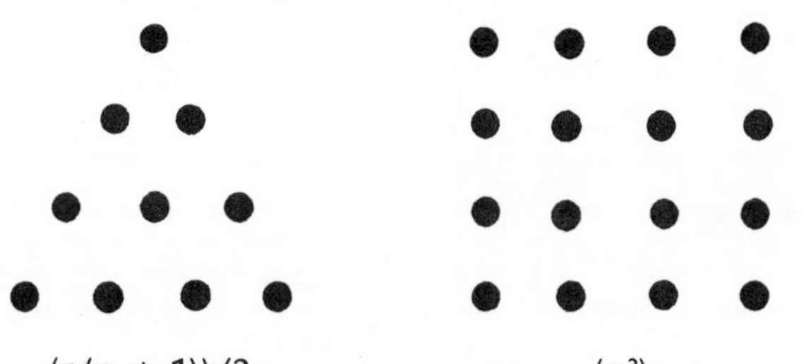

$(n(n + 1))/2$ (n^2)

Numbers which can represent a square pattern of dots are called *square numbers.*

Find out if there are numbers that are both triangular and square.

60. Find mutually distinct positive integers W, X, Y, and Z, each less than 15, such that $W^3 + X^3 = Y^3 + Z^3$.

61. Given a three-digit number, subtract the "inverted" three-digit number, and print the result. For example, 632 should give the result $632 - 236 = 396$.

62. Find those integers C, from 1 to 50, which can be written in the form $C^2 = A^2 + B^2$, with positive integers A, B.

63. The square of 5 is 25, and 5 is the last digit of 25. The square of 90625 is 8,212,890,625 and 90625 are the last five digits of 8,212,890,625. Find more numbers, less than 1000, which are the last digit or digits of their squares.

64. The sum of the first two digits of the integer 3025 plus the last two digits is 55. If you square 55 you obtain the original number: $30 + 25 = 55$, $55^2 = 3025$. Determine all numbers less than 10,000 with this property.

65. The three consecutive numbers: 72, 73, and 74 are unique since each equals the sum of 2 squares: $72 = 6^2 + 6^2$, $73 = 3^2 + 8^2$, $74 = 5^2 + 7^2$. Determine all similar sets of numbers less than 1000.

66. The number 20 can be written as the sum of two squares: $20 = 2^2 + 4^2$. Determine all numbers less than 100 which can be written as the sum of two squares.

67. Find every two-digit integer that equals the sum of the squares of its digits.

68. Four and nine are called square numbers because $2^2 = 4$ and $3^2 = 9$. Certain pairs of numbers when added or subtracted give a square number. For example,12 and 37: $12 + 37 = 49$ and $37 - 12 = 25$. Determine all the pairs of numbers less than 80 which give a square number when added and subtracted.

69. The late G. H. Hardy, a leading mathematician of his time, once rode in a taxi cab to see his ill Indian mathematician, Srinivasa Ramanujan. The number of the taxi cab was [?], and he remarked "That the number seemed to me rather a dull one, and that I hoped it was not an unfavorable omen. 'No,' he replied, 'it is a very interesting number; it is the smallest number expressible as the sum of two cubes in two different ways.' " [?] $= A^3 + B^3$ and $X^3 + Y^3$. Determine the taxi cab number.

70. For each of the following pairs of numbers, find two numbers so that the sum of your two is the first number in the given pair and the product is the

second number in the given pair: 42, 392; 75, 756; 94, 840; 296, 17679.

71. Using the integers one to nine and taking them three at a time, how many ways can the sum 15 be obtained?

72. Demonstrate that the sum of the squares of five consecutive integers is always divisible by 5.

73. Input an integer N and determine whether it is a perfect square.

74. The sum of the squares of the first N integers is given by the formula: N(N + 1)(2N + 1)/6. Use this formula to compute the sum of the squares of the first 150 integers.

75. Find all even natural numbers between 100 and 150 that are half the sum of their proper divisors.

76. Find the three smallest odd numbers which are each the sum of the squares of positive integers in two different ways.

77. Find the numbers A, B, C where their sum is 43 and the sum of their cubes is 17299.

78. Find the positive integers whose sum is 20 and whose product is a maximum.

79. Paradoxically, certain fractions may be reduced by striking out a common integer in the numerator and denominator. For example: 16/64 = 1/4, 154/253 = 14/13. Find all such fractions up to 998/999.

80. For each integer from 1 to 40, find the smallest square that begins with the digits of that integer.

81. An integer X, when divided by the numbers 2 through 12 always leaves a remainder of 1. X is also divisible by 13. Determine the smallest integer that fits this description.

82. A certain number minus the sum of its digits equals 36. Also the sum of the digits plus the product of the digits equals the number minus eight. Determine the number.

83. Find a sequence of 7 consecutive integers none of which is a prime number.

84. Find the squares of all integers through 100 whose tens' digit is odd.
85. Determine the sum of all integers between 100 and 1000 that are divisible by 14.
86. Print all numbers up to 1000 which are not divisible by any integer less than 10.

CHAPTER 8

SCIENCE, CHEMISTRY, AND PHYSICS

Contained in this chapter are problems related to science, chemistry, physics, biology, and engineering.

1. Convert an angle from degrees, minutes, and seconds to degrees with a decimal fraction.
2. Find the distance light travels in x years.
3. Light can travel 300,000 kilometers per second. Calculate the distance light can travel in one minute.
4. Read the height from which an object falls and determine the time it will take to fall.
5. Write a program which will print grams/mole and atoms/mole from atomic weights and molecular formulas.
6. Temperature is a measure of the concentration or intensity of heat energy in a body. There are four scales to measure temperature: Fahrenheit (F), Celsius (C), Kelvin (A), and Rankine (R). Many laboratory experiments involve converting from one scale to another. One example would be to express Fahrenheit readings in Celsius, Kelvin and Rankine. to convert Fahrenheit temperatures to Celsius, subtract 32° from the Fahrenheit reading and multiply the difference by 5/9. The product will be the Celsius equivalent. To

convert Celsius temperatures to Kelvin, add 273° to the Celsius reading. To convert Fahrenheit temperatures to Rankine, add 460° to the Fahrenheit reading. Write a program to convert 30 Fahrenheit values to their equivalent values in Celsius, Kelvin and Rankine.

7. Convert Celsius temperatures to Fahrenheit.

8. Convert Kelvin temperatures to Fahrenheit and Celsius.

9. A Chinese business man in Chicago wants to set his hotel room thermostat for 26° Celsius but the thermostat is marked off in a Fahrenheit scale. Write a program to make the conversion.

10. Generate two temperature conversion tables and print them. One table should give the Celsius equivalent for the Fahrenheit temperatures from 50° to 100° every 5°. That is, the Fahrenheit values should be 50°, 55°, ..., 100°. The other table should give the Fahrenheit equivalent for Celsius temperatures from 0° to 20° every 2°. That is, the Celsius values should be 0°, 2°, ..., 20°.

11. Mary Winslow and Don Carver set out to create their own thermometers. Mary calls the freezing point of water on her scale 40 degrees, while Don calls his freezing point 25 degrees. Mary makes the boiling point of water 280 degrees and Don, 125 degrees. Convert 12 values on Don's thermometer to equivalent values on Mary's thermometer.

12. Input the candle power of a light source, P, and its distance, D, from a point. Determine the illumination on an object placed at that point.

13. The drag and lift forces on a rocket can be approximated by the equations Drag = D d AV^2 and Lift = L d AV^2. D and L are the experimentally determined drag coefficient and lift coefficient, respectively; d is the density of air; A is the cross-sectional area of the rocket; and V is its velocity. Input several different values of D, L, d, A and V, and print the values of the drag and lift forces.

14. Given the atomic weight of an element, use the equation $E = mc^2$ to find the amount of energy

produced when that element is converted to energy.

15. Given the height of an object, the object distance, the image distance, and the focal distance, find the height of the image.
16. Input the angle of refraction of a light ray and find the angle of incidence for any given index of refraction.
17. The number of chirps that a cricket makes in a minute is a function of the temperature. As a result, it is possible to tell how warm it is by using a cricket as a thermometer! A formula for the function is t = (n + 40)/4. t represents the temperature in degrees Fahrenheit and n represents the number of cricket chirps in one minute. Determine and output values for t for n equal to 40, 50, 60, 70, ..., 140, 150.
18. The braking distance in Celsius, C, of an experimental car has been found to be 1.8 times the square of the car's speed s^2; that is, $C = 1.8s^2$. Compute and print the braking distance for speeds from 40 kilometers per hour to 160 kilometers per hour.
19. The gravitational attraction between any two bodies in the universe is given by the formula $F = (G \times M \times N)/R^2$ where M and N are masses of the bodies (in kilograms), R is the distance between them (in meters), and G is the constant of gravitation. Establish values for M, N, R and G, and compute and print F.
20. The amount of time it takes an object to travel a certain distance when traveling at a constant rate of speed is given by the formula t = d/r where d is the distance and r is the rate of speed. Compute t when r = 40 kilometers per second and d varies from 10 meters to 200 meters in increments of 10 meters.
21. The average speed V of a molecule of gas at a temperature T is given by the formula $V = \sqrt{3KT/M}$ where T = temperature (in degrees Kelvin), M = mass of the molecule (kilograms) and K = Boltzmann's constant. Establish the input values and compute V.
22. In the 17th century, Galileo Galilei supposedly dropped a pair of stones from the top of the Leaning Tower of Pisa in an attempt to demonstrate that heavy

stones and light stones fall at the same rate. Assuming that $g = 32$, and $t = 3$ seconds, use the equation $d = 1/2gt^2$ to write a program for determining how far the two stones will fall in the first 3 seconds.

23. A woman drops her pocketbook from the top of the Sears Tower (1451 feet high). Write a program for determining the impact velocity at the street level. Use the formula $V = \sqrt{2gh}$, where h is the height of the Sears Tower and $g = 32$ ft/sec^2 is the earth's gravitational constant.

24. A ball is dropped from a height of 10 meters. It bounces back each time to a height two-thirds of the height of the last bounce. Write a program to determine approximately how far the ball will have traveled when it hits the surface on the twentieth bounce.

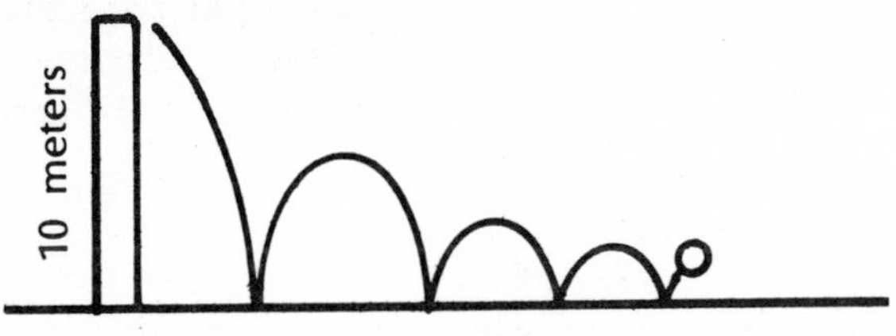

25. A ball falls from a height of 70 meters and rebounds 5/8 of the height from which it falls. Compute and print the rebound height for each of the first 30 times the ball hits the surface.

26. A ball is dropped straight down from a tower 86 meters high and rebounds each time to 32% of its previous height. Determine the total vetical distance traveled by the ball before it comes to rest.

27. If a ball bounces 0.8 the distance it is dropped, how many bounces will a ball make before it rises to less than 1 foot? If it is dropped from 6 feet.

28. If a body falls under the action of gravity (make no allowance for air resistance) the distance traveled is given by the formula, $S = 16t^2$, where S is the distance in feet and t is the time in seconds. How far does a falling body drop during the sixth second?

29. The equation for determining the current flowing through an alternating circuit is

$$I = \frac{E}{\sqrt{R^2 + \left(2\pi FL - \frac{1}{2\pi FC}\right)^2}}$$

where I = current, amperes
E = voltage, volts
R = resistance, ohms
L = inductance, henrys
C = capacitance, farads
F = frequency, cycles per second

Compute the current for a number of equally-spaced values of capacitance for voltages of 1.0, 2.0 and 3.0. Input to the program should be values for R, F and L. The starting and terminating values of capacitance, as well as the increment, are also input to the program.

30. The formula for calculating distance is d = rt where d stands for distance, r stands for rate and t stands for time. Compute the distance for the following values of r and t: r = 45 kilometers per hour, t = 6 hours; r = 60 kilometers per hour, t = 2.8 hours; r = 125 kilometers per hour, t = 1.5 hours.

31. Determine the resultant of two forces which act simultaneously at the same point on a body, given the magnitude and direction angle of each force.

32. Find approximately the amount of work required to lift a 900 kilogram body to a height of 100 kilometers from the earth's surface. Take the force to be proportional to $1/x^2$, where x is the distance from the center of the earth, and take the earth's surface to be 7400 kilometers from the center.

33. A ball is thrown in the air with an initial velocity V at angle A with the horizontal at a height D above the ground. Plot the trajectory of the ball using n-second intervals until it hits the ground.

34. A ray of light traveling in the air at a speed of 344,472 kilometers per second enters the water in such a way that the angle of incidence is 45° and the angle of refraction is 30°. How fast is the light ray traveling in the water?

35. Compute the equivalent resistance between points A and B of the following circuit:

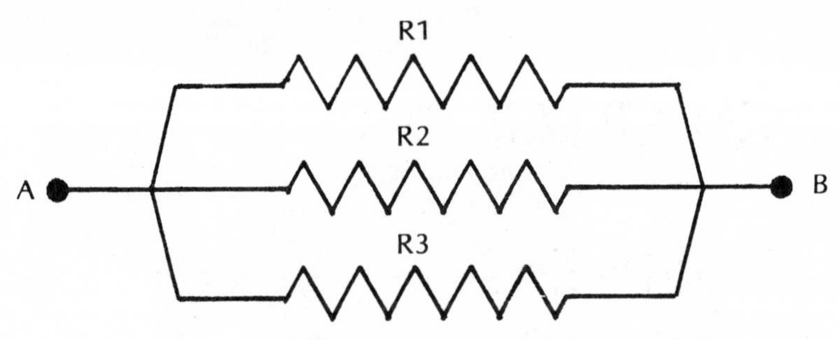

36. The increase in population of a bacteria culture with time is directly proportional to the size of the population. Thus, the larger the population, the faster the bacteria will increase in number. Mathematically the population at any time can be expressed as $P = R(1 + ct + ct^2/2 + ct^3/6 + ct^4/24 + \ldots + ct^n /n!)$. P = bacteria population at time t, R = bacteria population at the reference time, t = time in hours beyond a reference time, and $c = 0.0289$. Compute the population multiplication (P/R) at 2, 5, 10, 20, and 50 hours beyond the reference time. Use only the first 10 terms fo the series.

37. A small town in western Kansas devised an air pollution index such that less than 37 is "acceptable", from 37 to 55 is "unpleasant", and above 55 is "hazardous". Write a program that accepts an air pollution index as data and prints the appropriate description of the air.

38. There are 10 simple animals in a laboratory culture and enough food for 1000 such animals at time zero (the present time). Every hour, the population doubles, and enough food is added to the culture to feed 4000 more animals than at the previous hour. When, if ever,.will the population outgrow the food supply?

CHAPTER 9

BUSINESS

It is the purpose of this chapter to present a variety of simple introductory business problems. You will find problems related to payroll, interest, depreciation, inventory control, sorting, tax computations, investments, mortgages, and employee bonus calculations.

1. Sara Iverson sells bibles at $3.00 each plus $.65 for postage and handling. Calculate her total receipts for two weeks in which she sold 164 bibles.

2. The Penquin Press supplies schools and colleges with textbooks. It offers reduced rates on orders of 30 or more copies of the same textbook. A certain text is priced as follows:

   Under 30 copies — $6.95 per copy
   Thirty or more copies — $6.00 per copy

   Compute the cost for schools ordering the following number of texts:

   School A — 35 copies   School C — 70 copies
   School B — 12 copies   School D — 20 copies

3. The Northern Freight Company charges the following rates on merchandise shipped from Boston to Los Angeles: $75 per ton for the first 10 tons, $35 per ton for every ton over 10. How much would it cost to send shipments weighing 12 tons; 36 tons; 8 tons; 100 tons; 1260 tons?

4. The student population at Atlanta Technical School increases by 8 percent every year. If the current student population is 2400, how many students can this school expect in 10 years?
5. A customer ordered four books which retail at $8.95 and carry a 20% discount, three records at $3.50 with a 15% discount, and one record player for $59.95 on which there is no discount. In addition, there is a 2% discount allowed on the total order for prompt payment. Compute the amount of the order.
6. How many ways can change be made for 50 cents using quarters, dimes, nickels, and pennies?
7. Write a program that keeps track of your (or your dad's) checking account. It should add deposits, subtract the amounts of checks you write, subtract the check charge the bank makes, and print the balance at the end of the month.
8. Janet Helmich worked on four jobs each paying a different hourly rate. Determine the amount of money Janet earned in a week if she worked the following hours at the indicated rates: Job A — 12 hours at $3.20 per hour, Job B — 10 hours at $4.10 per hour, Job C — 8 hours at $3.80 per hour and Job D — 13 hours at $2.95 per hour.
9. Calculate a person's gross weekly pay, given his rate per hour, overtime rate per hour, number of hours worked that week, and number of overtime hours worked that week.
10. Assume that the withholding tax on a weekly salary is computed as follows: 15% of the difference between an employee's gross pay and $10 times the number of dependents he claims. Input the values for an employee's gross pay and the number of depenents; print the employee's withholding tax.
11. Job X lasts 30 days and pays $10 per day; job Y lasts 30 days and pays as follows: $1 first day, $2 second day, $3 third day, and so forth. Which job pays more?
12. Create an employee file for the Northern Ski Factory. The file should contain the employee's identification number and birthdate. The company's mandatory

retirement age is 65. Print the identification numbers of those employees who must retire within the next year.

13. Write a program in which you can input an employee's code (a 7-digit identification number), his hourly wage, rate, and the number of hours he worked during the week. The program should print each employee code, the salary due, and the word OVERTIME if the employee worked more than 40 hours.

14. Roger Bigstuff earns $4.50 per hour for up to 40 hours, and $6.75 for each hour worked in excess of 40 hours in a single week. He worked 53 hours last week. Calculate his weekly earnings.

15. The Southwest Lumber Company (SLC) pays a bonus to the employee who makes the most dog houses in a given year. The dollar size of the bonus is determined by multiplying by 10 the difference between the total number of doghouses made by the highest person and that made by the lowest person.

    Bonus = 10 × (Highest number − Lowest number)
    SLC has seven employees. Write a program to read the twelve monthly production figures for each person, calculate the bonus, and print the identification number of the employee who is to receive the bonus.

Monthly Production Figures

Employee	J	F	M	A	M	J	J	A	S	O	N	D
1	101	93	107	63	121	77	102	72	79	76	80	53
2	99	80	82	60	65	80	91	95	63	75	92	61
3	79	100	122	76	67	80	80	90	100	60	91	69
4	40	89	100	90	92	95	96	89	79	72	90	72
5	121	101	98	97	103	104	89	99	107	90	76	49
6	99	89	60	99	98	88	95	96	89	90	91	60
7	79	89	90	70	90	88	82	63	70	75	80	70

16. A customer has $3400 to spend and wants to purchase a lot having an area of at least 9000 square meters. Given data for twenty lots, write a program that will input the person's requirements and determine whether they can be met by any of the given lots. Any lot or lots found to meet the requirements should be printed in the following format:

IDENT	LENGTH	WIDTH	VALUE
102	100	110	2950

17. A formula for computing the property tax on a piece of real estate is T = (A/100)R where T = tax, A = assessed value, and R = rate per $100. Calculate the tax on a house worth $33,000 with a tax rate of $4.25 per $100.

18. The Metal Toy Company is reducing the price of its metal trucks and cars by 12%. Calculate the average reduction in price of the trucks and cars which originally sold for $8.95, $12.50, $5.50, $14.25, $9.50, $7.50, $10.00, and $3.20.

19. Simple interest calculations involve arithmetic progressions. If P is the principal placed at an interest rate i for a period of n years, the amount A at the end of n years may be found by using the formula A = P(1 + n). For example, if P = 2000, i is 7, and n is 10, then

$$\begin{aligned} A &= 2000(1 + (10 \times .07)) \\ &= 2000(1 + .7) \\ &= 2000(1.7) = \$3400 \end{aligned}$$

Input the following values: P = 5000, i = 6, n = 15. The program is to calculate the amount at the end of each period from 1 to 15 years, printing out values of the period and the amount.

20. One thousand dollars was deposited with a Savings and Loan Company that pays 5 ¾ % interest, compounded quarterly. This deposit was made on January 1, 1970. On January 1, 1971, an additional deposit was made in the amount of $500. This was repeated on January first of each year until the final deposit was made on January 1, 1979. (a) How much is in the account on January 1, 1980, and (b) How much of this amount is interest?

21. Ann Smarts finds a $15 skirt that she simply must have. There is a slight problem — she doesn't have fifteen dollars. She does have credit at the store, however, and all the ads say "easy credit". Why not? The store charges Ann just $1.25 down which she must pay when she purchases the skirt. The remaining money will be paid at the rate of $1.25 per week for fifteen weeks. Compute the simple annual interest rate.

22. Manhattan Island was purchased in 1626 for $24. If

those early investors had decided to invest the same amount at 7% interest, compounded quarterly, how much would their investment be worth today?

23. Simple interest is paid on $300 invested at r% for n years. Write a program to print an interest table for values of r from .05% to 6.5% and for integral values of n between 1 and 25.

24. The compound interest formula is $A = P(1 + I/100)^N$ where P is the principal (the amount originally invested or deposited), I is the yearly rate of interest, N is the number of years, and A is the amount (principal plus interest). Using an initial deposit of $2000 invested at 5% for 5 to 20 years, compute a table showing the interest for each year.

25. How much money will you have in the bank if you deposit $10 at the beginning of every month for 20 years in a savings account which pays 5½% interest compounded monthly?

26. A father wants to put enough money in a college savings account that pays 6% interest compounded daily so that his son, who was just born, will have $12,000 when he is 19 years old. Determine the amount he should put into the account.

27. Suppose that you borrow $2000 at 9% interest per year for five years. How much "rent" (interest) must be paid?

28. You have $100. If you invest it at 6% interest compounded quarterly, how many years will it take before you have $50,000?

29. Suppose you want to be a millionaire by the time you are 50 years old. How much would you have to put into a savings account that earns 6% interest compounded quarterly?

30. Compare $150 compounded monthly at 5½% interest, quarterly at 5¼%, and daily at 5%.

31. A $23,000 mortgage is to be repaid at the rate of $260 per month. The interest is charged at the rate of 6% each year, calculated each month. The program should be designed to generate a four column table which will show the payment number, the balance, the

interest for each month, and the amount paid on the principal for each month. The new balance for each month is obtained by subtracting the amount paid on the principal from the old balance.

32. A dentist wishes to solve a simple accounting problem. An x-ray machine worth $12,000 is to be depreciated over 20 years by the use of a double declining-balance depreciation. Produce a printout of the depreciation and the book value of the x-ray machine for the 20 year period.

33. If you inherited $1,200,000 (a dream, perhaps!) and had it invested with an average return of 6.2% per year, what would your annual income be?

34. You have just invested $100 in the Western Investment Company. The Board of Directors guarantees that you will double your investment every two years. Compute and print a table showing your investment for 30 years. The program should print a table similar to the following:

    2 YEARS — $200
    4 YEARS — $400
    6 YEARS — $800

35. The Shifty Loan Company loans money at 12.5% interest. However, if the loan is $1000 or more the interest rate is reduced to 10%. Find the interest charged for one year on loans of $500, $2200, $3000, $250, $10,000, $375.

36. Tom's father invested $2000 in a trust fund for him on the day he was born. The fund earns 7% interest compounded weekly. Write a program that will determine what the fund will be worth when Tom is 30, 40, 55, 65, 85, and 91 years old.

37. A man and his wife plan to save money from their paychecks to make a down payment on a home. They intend to make monthly deposits of $100 in a savings account on which the bank pays monthly interest at the rate of ½ of 1%. How many months will it take to accumulate $5000?

38. Suppose that $150 is invested at the beginning of each year for 15 years and that the money earns interest at the rate of 9 % per year, payable at the end of each

year. If the interest is reinvested, how much money will have accumulated by the end of the fifteenth year?

39. Write a program which will compute the total interest earned given the total invested, interest rate, number of years invested, and the number of times per year interest is paid.

40. A Georgia peanut grower wants to invest his $18,000 inheritance in such a way that he can double his money every 7 years. Write a program to determine what compound interest rate he would have to get to achieve his rather ambitious goal.

41. Mary's father invested $1000 in a trust fund for her on the day she was born. The fund earned 10% interest compounded yearly. Mary forgot all about this trust until she was 56 years old. How much was the fund worth then?

42. Sherrie has $300 she can put in a bank for 1 year. The First National Bank pays 6% interest compounded annually. The Second National Bank pays 6% interest compounded quarterly, the Southern Savings and Loan pays 6% interest compounded weekly, and the Moon Bank pays 6% interest compounded daily. How much more money would Sherrie make by getting the interest compounded daily instead of yearly? Find out how much $300 would be worth in each of the four banks at the end of one year.

43. Bill Williams plans to borrow $150 and wants to compare various loan plans. Suppose that the Local Loan Company will lend him money at compound interest at 1% monthly, and the Commercial Loan Company will lend him the money at simple interest at 1⅛% per month. Compare how much Bill would have to pay under each plan after 12, 24, 36, or 48 months.

44. Wilma Peterson intends to deposit $1000 in a savings account and leave it there for 10 years, accumulating interest. Bank A compounds the interest in its savings accounts annually, bank B compounds it every 6 months, bank C every 3 months, and bank D every month. Banks A and B pay interest rates of 6.25%, bank C pays 6%, and bank D pays 5.75%. In which bank should Wilma deposit her money?

45. A bricklayer borrows $1000 and agrees to pay back $90 at the end of each month for 12 months. He therefore pays back $1080 altogether, which might appear to represent interest of only 8% on the original loan. By making calculations with a few different rates, compounded monthly, determine the true annual interest rate to within .1%.
46. The Third National Bank pays interest at an annual rate of 5% on accounts with less than $200, 6% on accounts with $200 to $1000, and 7% on accounts with more than $1000. Compute the interest on an account whose balance is supplied as data.
47. Set up a table of amounts that $100 will be at the end of 10, 15, 20, and 25 years at 5%, 5½%, 6%, 6½%, and 7% interest per year compounded monthly. Print the years across the top and the rates in the first column of each row.
48. A $35,000 mortgage is to be repaid at the rate of $310 per month. The interest is charged at a rate of 8% each year, calculated each month. Write a program to compute a four-column table which will show the payment number, the balance, the interest for each month, and the amount paid on the principal for each month.
49. A mortgage of $3000 is to be repaid in equal monthly installments over a given period of years. The interest is calculated as a percentage of the loan outstanding at the time of each repayment. Print a table of monthly repayment charges for periods ranging from 1 to 25 years at rates of interest from 3% to 10%. It can be shown that the monthly repayment on a loan of $P over a period of n years is given by $(Py^{12n}(1 - y))/(1 - y^{12n})$ where y is the growth rate per month. For example, if the interest 7% then y = (1 + 7/1200).
50. Many banks compute interest quarterly; that is, every three months. Show the difference between investing $200 at 6.0% compounded quarterly and investing $200 at 6.25% compounded annually, over a period of 25 years.
51. The Nelson family is buying a house with a mortgage from the Security Savings Bank. The original loan was

for $45,000 at 8% interest per year. Mr. Nelson makes a monthly payment of $375, which includes the interest and payment toward the loan. The interest is computed each month on the "unpaid balance" of the loan. The first month, the interest is $i = (45{,}000) \times (.08) \times 1/12 = 300$. Therefore, the amount of the monthly payment toward the loan is $375 – $300 = $75. The unpaid balance becomes $45,000 – $75 = $44,925. The interest for the second month is calculated on this new balance. Print a payment table for the first 72 months.

52. Mary purchases her weekly grocery supply at the supermarket. Let X be the total value of her purchases and Y the amount of money she gives the clerk. Calculate the number of bills and the number of coins of each denomination she receives in change (difference between X and Y).

53. Jack Wilson has $7600 to spend and wants to purchase a lot having an area of at least 12,000 square feet. Make up data for 30 lots and write a program to input Mr. Wilson's requirements and determine whether they can be met by any of the given lots. Print out the identification, size, and price of all lots that meet Mr. Wilson's lot requirements.

54. Write a program to find marginal revenue, marginal cost, and total profit, given total revenue and total cost at each level of output.

55. Input the following table, sort the entries by key, and print a table of values in key order.

Key	Value
2	14
6	301
32	1632
4	171
11	6321
1	148
15	7
9	23
25	666
17	31

56. Write a program that prints monthly bills for CHARGEALL, a credit card company. It should add payments made in the past month, subtract the cost of purchases made, and subtract a 1.5% monthly finance charge on the unpaid balance.

57. Write a program to drill a student in percentage problems.

58. The Northern Fabric Company uses 3 yards of cloth to make each of 3 types (A, B, C) of woman's dresses. The following table indicates percentages of dacron, cotton, and wool in the three dress categories.

	Dress A	Dress B	Dress C
Dacron	40	45	25
Cotton	10	30	15
Wool	50	25	60

Find the number of yards of cloth (Dacron, Cotton, and Wool) needed to make 100, 200, 300 coats if the ratios of A to B to C coats is always 1 to 3 to 6.

59. A department store has a clothing item in stock, classified as small, medium, and large in colors red, white, and yellow. The stock on hand (by size and color) is given in Table A where rows represent size and columns represent color. The price per unit is also given in the same classification scheme in Table B. How much inventory by size and color does the merchant have on hand in dollars?

Table A

	R	W	Y
S	6	14	2
M	12	22	8
L	7	3	4

Table B

	R	W	Y
S	3.95	3.95	4.25
M	4.95	4.95	5.25
L	5.45	5.95	6.50

60. Mr. Wilson is offered employment by the Southern Manufacturing Plant and is afforded the opportunity of taking two different methods of payment. He can receive a monthly wage of $500 and a $5 raise each month, or he can receive a monthly wage of $500 with a yearly raise of $80. Write a program to determine the monthly wages for the next 10 years in each case. The program should determine the cumulative wages after each month, and from the information determine which is the better method of payment.

61. At a price of $1.00, a pipe dealer can sell 1000 pipes that cost him 50 cents each. For each cent that he lowers the price he can increase the number sold by 50. What price will maximize the profits? Hint: Profits = Sales - Cost. Clearly, the right price lies somewhere between $.50 and $1.00. Calculate the profits at each of the prices in this range and choose the one which yields the maximum profit.

62. The Pan American National Bank requires a monthly report, which contains the following information:
    - The number of overdrawn accounts.
    - A list of the accounts overdrawn.
    - The number of accounts with balances in excess of $100 but less than $1000.
    - A list of those accounts whose monthly balance exceeds $1000.
    - The number of such accounts.
    - The amount of the cash balance of the largest depositor.
    - The cash balance of the smallest depositor.

    Create a program of Account Balances and list the above information.

63. Write a program to produce a facsimile of a check. The program should input the payee, data, signator, bank name, bank address, transit numbers, check number, and amount of the check.

64. The Goodwill Card Company decides to give each employee a bonus of one-eighth of her or his annual salary. Find the bonuses paid by this company if the annual salaries of their seven employees were $12,000, $14,000, $9,000, $6,500, $7,800, $10,400, $8,200.

CHAPTER 10

FUN AND GAMES WITH THE COMPUTER

Up until the invention of the computer, game playing was primarily restricted to mere humans or special purpose machines. Today, people all over the world are spending a considerable amount of time programming computers to play games. Why? One reason is that game programs provide us with excellent opportunities for learning how to solve problems with computers. The beginning computer user can understand the problem to be programmed in a minimum amount of time; therefore, he can devote more time to learning about the computer, algorithm development, programming languages, and techniques of problem solving with a computer.

Another reason why humans are using computers to play games is that games are often good analogies to actual situations involving humans and their environment. Gaming is being applied to business management, scientific experiments, and military war games. Business executives are playing games with digital computers that simulate the operation of their business. Games of this type allow the executive to keep an active study of his employees, to learn more about his company, and to simulate all activities of his company.

Contained in this chapter are games, puzzles, and mathematical recreations that can be programmed for play on a computer.

1. Write a program that simulates the throwing of two dice.
2. Draw five cards at random from a 52-card deck and print the suit and value of each card.
3. Deal and analyze a poker hand.
4. Write a program that will deal bridge hands.
5. Program the computer to "draw" a picture on the terminal.
6. The computer tries to guess a number you have in in mind. First it guesses a number and you tell it if the number is too high, too low, or correct. On the basis of the information you give, the computer guesses again. This continues until the computer guesses the number. Write a program to play this guessing game.
7. Write a program that asks two players to guess which number between 1 and 75 the computer randomly picked. The program should give 15 points to the player who was closer.
8. Place five coins on five squares so that no two coins are in the same row, column, or along a diagonal.

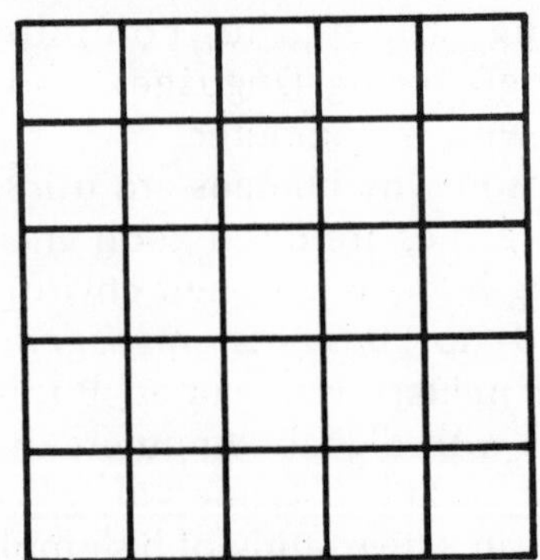

9. Simulate 100 times a game between Tom and Laura both of whom have 20 pennies. A coin is tossed. If it comes down heads, Tom wins a coin from Laura; if it comes down tails, Laura wins a coin from Tom. On the average, how many tosses will be needed before either Tom or Laura loses all their coins?

10. Input the number of pins John Wilson knocked down with each ball in each frame and compute his bowling score.

11. Solve the following problem. Substitute numbers for the letters.

$$\begin{array}{r} \text{HALF} \\ +\text{HALF} \\ \hline \text{WHOLE} \end{array}$$

12. Jack was taking his pet fox, a sack of corn, and a prize goose to the county fair. He came to a narrow footbridge over a river across which he could carry only one of his belongings at a time. If he left the goose and the corn alone, the goose would eat the corn. If he left the fox and the goose alone, the fox would eat the goose. How did he manage to get them all to the other side of the river?

13. Do you think a computer could be used to solve a recreational logic problem such as the following? If so, write a program to do so.

    The census taker, while taking head counts in a nearby village, questioned the one-armed owner of a small rundown shack. He pointed to another one-armed gent who was asleep. "Who is he?", he asked. The one-armed man replied, "Brothers and sisters have I none, but that man's father is my father's son." The object of the puzzle is to determine who the sleeping man is.

14. Three sailors, shipwrecked with a monkey on a desert island, gathered in one day a pile of coconuts that are to be divided early the next day. Sometime during the night, one sailor arises, divides the pile into three equal parts, and finds one coconut left over, which he gives to the monkey. He then hides his share. Later during the same night, each of the other two sailors arises separately and repeats the performance of the first sailor. In the morning all three sailors arise, divide the pile into three equal shares, and find one lett over, which they give to the monkey. Compute how many coconuts were in the original pile. Since there is more than one correct answer, the program should consider coconut piles only in the range of 1 to 1000. One

answer is 79 and may be used to check the correctness of the program.

15. How many chess queens are needed to "cover" a chessboard? (A board is called covered when every square is either occupied or threatened). Shown are four queens covering a 6 × 6 board. Determine by exhaustive search whether or not three queens could do the same job.

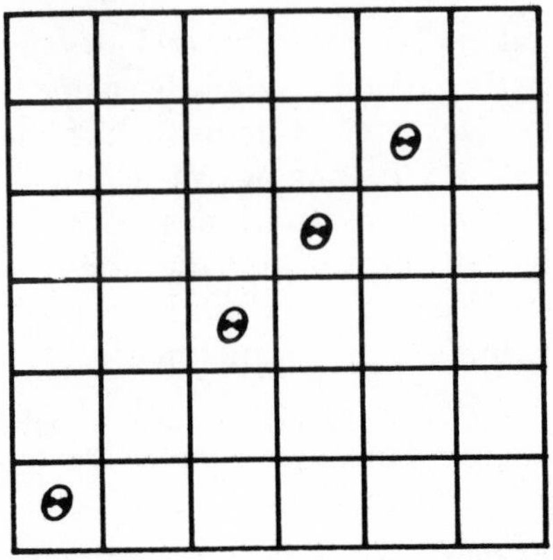

16. A game is played by rolling a die. If the outcome is an even number (2, 4, 6), the player receives an amount equal to the number on the die. If it is odd (1, 3, 5), he loses an amount equal to the number on the die. Simulate this game on a computer.

17. Write a program that deals a hand of bridge, then makes an opening bid.

18. Ten poker hands of five cards each can be dealt from a deck of 52 cards. Simulate this process a few hundred times, then find and output the number of poker hands that appeared with four of a kind, a full house, three of a kind, two pairs, and a single pair.

19. Simulate 10 hands of a two-person game of "Five-Card-Showdown." The game is played with a standard deck of 52 playing cards with four suits, deuce through ace. For purposes of the game, suits are irrelevant and aces are high; deuces, low. Winning combinations are pairs, three of a kind, and four of a kind. If neither player has a pair or better, high card wins. If both players have the same high card, the game is declared a draw and both players' bets are returned. Also, if

both players have the same pair (or better), the game is declared a draw and their bets are returned. Each player places a $5 bet. The dealer deals five cards to each player, and the players match their cards to declare either a win or draw. The program should deal five cards to each player, compare the dealt hands, and tabulate how much each player has won or lost.

20. The "Poker Dice" game uses five dice, each marked from Ace to 9. The object of the game is to roll the dice and make the best Poker hand, in one, two, or three rolls. Any number of players may play the game. The first player tosses the five dice. He may accept the toss, or put any of the dice aside and toss the others again. After his second toss, he may put aside any of the dice, adding them to the ones previously put aside, and toss the remainig die or dice a third time. The first player's hand is now recorded and the next player tosses the dice in a similar manner. Betting usually consists of each player putting a chip in a common pot prior to tossing the dice. The player with the highest hand wins the pot. Simulate this game on the computer.

21. The "Under and Over Seven" game is found in small gambling joints rather than plush gambling casinos. In this game, two dice are thrown, and players bet that the total will be Over 7, Under 7, or Equal to 7. A player places his bet on a layout, and after the roll of the dice, is either paid at 3 to 1, even money, or his bet is lost. A bet on Over 7 wins when the roll of the dice produced a sum of 8, 9, 10, 11, or 12. An Under 7 bet would be a winner if the dice total was 2, 3, 4, 5, or 6. Both the Over 7 and Under 7 bets are paid off at even money. If the player puts a bet in the center of the layout and the dice total was 7, he would be paid off at odds of 3 to 1. Simulate this game on the computer.

UNDER 7	7	OVER 7

22. Dara is a board game played by the Dakarkari people of Nigeria, North Africa. The board consists of 30 depressions made in a board. Each player has 12 stones and places them one at a time in the holes in alternate turns of play. This play continues until both players have all their stones on the board. The players then alternately move a stone orthogonally to the next hole. The object of the game is to form three stones in a line in consecutive holes on the board. Note that three diagonally arranged stones do not count. Whenever a player obtains three stones in a row or column, he may remove one of his opponent's stones from the board. The game terminates when one player is unable to obtain a three-stone line.

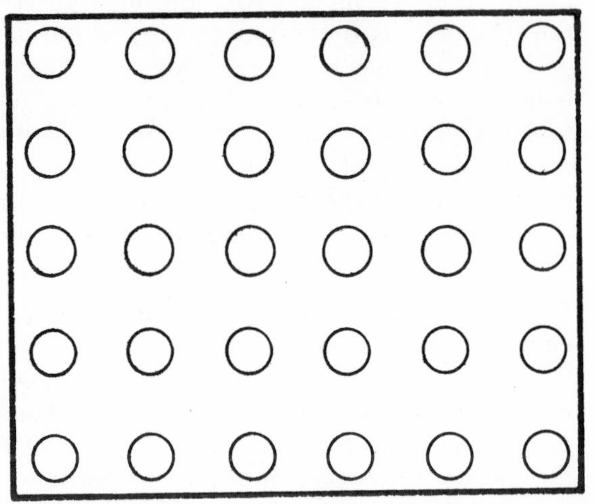

23. "Beat the Dealer" is a banking game played with a pair of dice. The game is started by the banker throwing the dice from a dice cup. The value the banker throws is marked on a layout with a poker chip. Now it is the player's turn to throw the dice. The player can win the game only if he throws a higher value than that marked on the layout. The banker wins on all ties.

24. The "marked square" game is played on a board with 9 rows and 9 columns. Players alternate in marking a square at a time, each using his own distinguishing mark. Whenever a player marks the last square in a horizontal, vertical, or diagonal line, he is credited with all the squares in this line. Each square is scored as

one point. The player with the higher final score wins the game. Simulate this game on the computer.

25. The King's power of movement on the chessboard is very limited. He can move only one square at a time. He can go into any of the squares — front, back, or side — adjacent to the square on which he stands. To complete a King's Tour we must move the King successively to every cell on the chessboard. An interesting thing about the tour shown is that the numbers indicating the path form a "magic square". Write a program to generate a King's Tour.

61	62	63	64	1	2	3	4
60	11	58	57	8	7	54	5
12	59	10	9	56	55	6	53
13	14	15	16	49	50	51	52
20	19	18	17	48	47	46	45
21	38	23	24	41	42	27	44
37	22	39	40	25	26	43	28
36	35	34	33	32	31	30	29

26. In a game of Bingo, 75 tokens numbered 1 to 75 inclusive are placed in a container that can usually be rotated or shaken to mix up the tokens. The caller in the game removes one token at a time and calls out its number. If the number is between 1 and 15 inclusive he will usually say "Under the B", and state the number. Similarly, the numbers between 16 and 30 are under the I, between 31 and 45 under the N, between 46 and 60 under the G and between 61 and 75 under the O. A bingo card is divided into 25 squares. The center square contains a free play. The other 24 squares on the card contain numbers in the range 1 through 75. The object of the game is to get 5 numbers in a row, column, or diagonal. Write a program to play the game of Bingo.

27. The WHEEL OF FORTUNE is a giant wheel with a diameter of about five feet. The rim of the wheel is divided into 60 sections. In 58 of the sections is paper money in denominations of $1, $2, $5, $10, and $20. The remaining two sections contain a Joker and a Flag. The Wheel of Fortune layout, which consists of seven corresponding numbers and symbols, is used by the players for placing bets. The wheel is spun and players bet that it will come to rest with the pointer at a specified money denomination. A player will win even money if he bet on $1 and the pointer stopped at the $1 bill. He will win $2 if he bet on $2 and the pointer stopped at the $2 bill. If the wheel stops at the $5 bill, the player will collect $5 if he bet on that value. If the wheel stops at the $10 bill, the player will win $10 if he has bet on that denomination. If the wheel stops on the $20 bill, the player will win $20 if he was betting on that value. The Joker and Flag pay-off at 40 to 1 odds and a player betting on this value would collect $40 if the wheel stopped on either one. On most wheels there are 22 $1 bills, 14 $2 bills, 7 $5 bills, 3 $10 bills, 2 $20 bills, 1 Joker, and 1 Flag. Write a program to simulate play on the Wheel of Fortune.

28. Chuck-a-Luck is a gambling game often played at gambling casinos. A player may bet on any one of the numbers 1, 2, 3, 4, 5, 6. Three dice are rolled. If the player's number appears on one, two, or three of the dice, he receives respectively one, two or three times his original stake plus his own money back; otherwise he loses his stake. Simulate this game on a computer.

29. This problem is provided for students who wish to "gamble" with their computers. The computer's typewriter (or CRT keyboard/display device) is used as the "slot machine", the RETURN key is used as the "handle", and instead of figures showing behind the "windows", the typewriter outputs three words representing the figures. There are 6 "figures" and the pay-offs vary approximately the same as the slot machines at Las Vegas. The coin size of the simulated slot machine is 25 cents. The figures are chosen in random order. Write a program to simulate the operation of this "slot machine" and print the money won or lost by the gambler after each simulated pull of the handle.

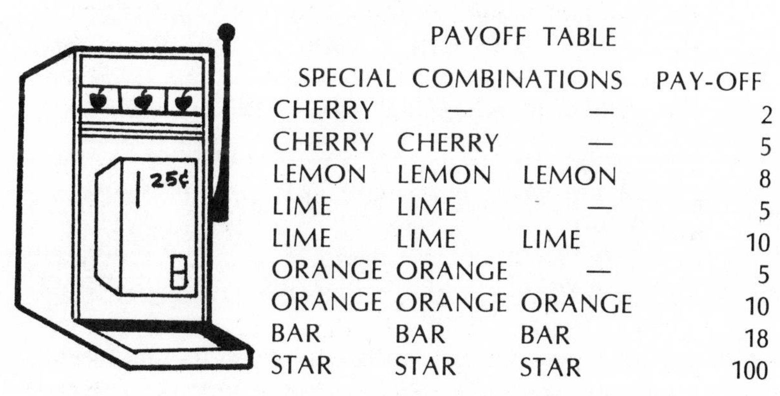

PAYOFF TABLE

SPECIAL	COMBINATIONS		PAY-OFF
CHERRY	—	—	2
CHERRY	CHERRY	—	5
LEMON	LEMON	LEMON	8
LIME	LIME	—	5
LIME	LIME	LIME	10
ORANGE	ORANGE	—	5
ORANGE	ORANGE	ORANGE	10
BAR	BAR	BAR	18
STAR	STAR	STAR	100

30. The game of craps, played with two dice, is one of the most popular gambling games. Only totals for the two dice count. The player throws the dice and wins at once if the total for the first throw is 7 or 11, loses at once if it is 2, 3, or 12. Any other throw is called his "point".If the first throw is a point, the player throws the dice repeatedly until he either wins by throwing his point again or loses by throwing 7. Simulate the game of craps on a computer.

31. Program the computer to play blackjack against human players with the computer being the dealer.

32. At Las Vegas, a man with $25 needs $50 but he is too embarrassed to wire his wife for money. He decides to invest in roulette and is considering two strategies: bet the $25 on the "red" all at once and quit if he wins or loses, or bet on "red" one dollar at a time until he has won or lost $25. Compare the merits of the strategies.

33. The game of NIM begins with three piles of counters. The players (one of which can be a computer) take turns removing counters and the player who picks up the last counter loses. Any number of counters can be picked up in each turn, as long as at least one counter is picked up and all are chosen from the same pile. Write a program to play Nim.

34. The game of markers begins with 15 markers in a row. There are two players who take alternate turns. On each turn a player may remove 1, 2, or 3 markers. Whoever takes the last marker wins. Write a program that allows a person to play the game of markers against the computer.

35. Write a program to simulate a mouse finding his way through a maze to some cheese. The maze will be a 20×20 grid. The mouse will start in the northeast corner of the grid, and the cheese is located in the southwest corner of the grid. The mouse may move one square at a time either north, south, east, or west, but not off the grid or to a square he has been on before. His movements will be determined randomly by generating random integers in the range 1-4. The program should print the positions of the mouse as he travels through the maze.

36. Determine the minimum number of moves that are necessary to make the white and black knights change places on the board shown. A knight moves one square diagonally in any direction, then one square to either side of the direction of the diagonal. The knight can leap over any piece on the board. (Note: This exchange can be accomplished with 16 individual moves).

37. In the game known as Yahtzee, five dice are thrown at one time. "Yahtzee" consists of having all the dice turn up with the same number. Simulate this game and print YAHTZEE everytime all the dice are equal.

38. Five Field Kono is played using the following layout:

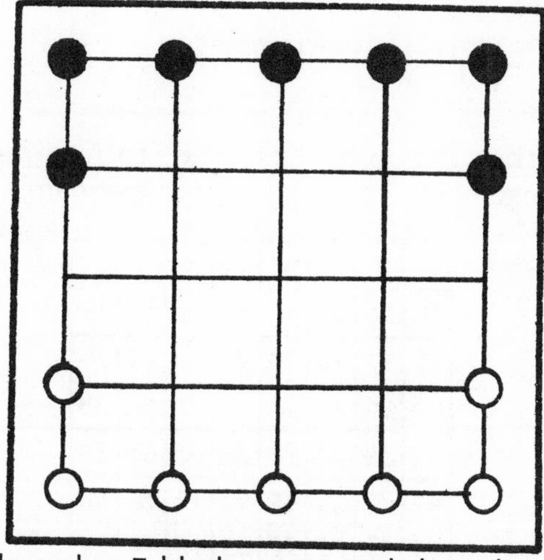

One player has 7 black stones and the other 7 white stones. The player with black stones always makes the first move. The players move one piece at a time, in alternate plays either backward, forward, or diagonally across the squares. The object of the game is to get the pieces across to the other side in the place of those pieces of the other player. The player who does this first wins the game. Write a program to play this game.

39. Magic squares are one of the oldest and most fascinating of all number curiosities. A *magic square* is an array of numbers arranged in the form of a square, so that the sum in every column, in every row, and in both main diagonals is identical. No number may be used twice in constructing the array. Write a program to generate an Order 5 magic square, consisting of 25 numbers.

17	24	1	8	15
23	5	7	14	16
4	6	13	20	22
10	12	19	21	3
11	18	25	2	9

40. Determine whether the following number arrangement represents a magic square.

64	2	3	61	60	6	7	57
9	55	54	12	13	51	50	16
17	47	46	20	21	43	42	24
40	26	27	37	36	30	31	33
32	34	35	29	28	38	39	25
41	23	22	44	45	19	18	48
49	15	14	52	53	11	10	56
8	58	59	5	4	62	63	1

41. Tac Tic, a game for two, is a variation of the game of Nim, which was invented by Piet Hein of Denmark. It is an interesting one to present because it has not yet been completely analyzed. Arrange 16 coins as in the figure. These are numbered here for ease of reference.

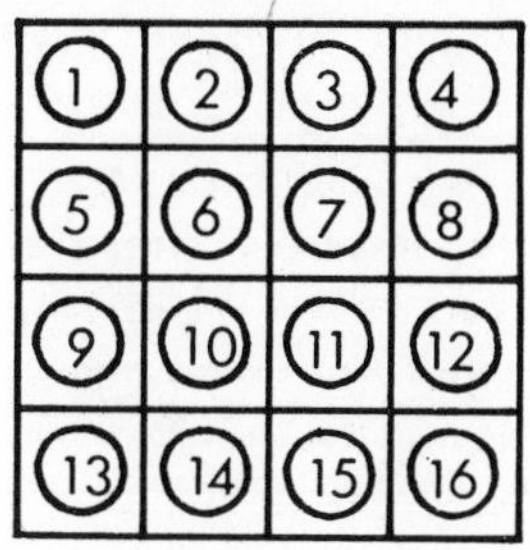

Players alternate removing any number of chips from any single row or column. However, as an additional constraint, only adjacent coins may be removed. For example, if player A removes coins 14 and 15 on his first move, player B may not take 13 and 16 in one move. The player who is forced to take the last coin is the loser.

42. In the Tower of Hanoi game, one must transfer a set of disks from one peg to another with a restriction that a larger disk must never be placed atop a smaller one. The number of movements required can be shown to equal $2^n - 1$ where n is the number of disks in use. A San Juan shoplifter is sentenced to stay in jail until she has transferred the disks on a Tower of Hanoi from one stake to another in accordance with the rules of the game. Write a program to determine how many separate moves she will have to make. There are 20 disks and 3 stakes.

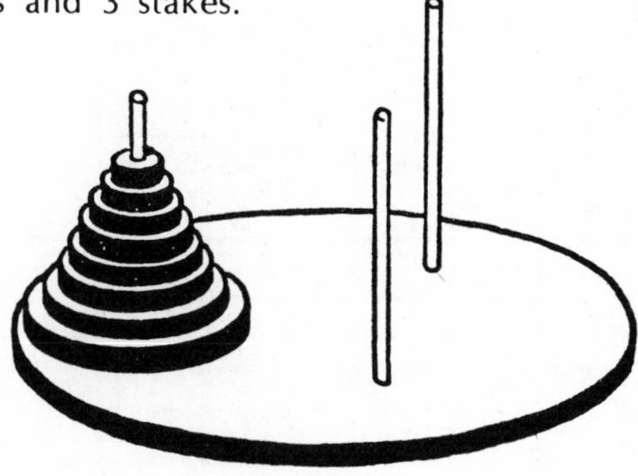

43. Write a program to play tic-tac-toe. The game is played with two players (one is a computer), alternately selecting squares of an Order 3 array. Print the board configuration after each move. When a player obtains 3 across, 3 down, or 3 diagonally, print the winner. Print an appropriate message if the game ends in a draw.

44. A pentomino is a plane figure formed by five contiguous equal squares. There are 12 possible ways to arrange five squares in this manner; therefore, there are 12 different pentominoes. The pentomino game is played by arranging the 12 pentominoes into a 6 × 10 rectangular box. Write a program to arrange several patterns of the pentominoes (there are over 2000 different patterns).

The 12 Pentominoes

45. A popular puzzle called *Instant Insanity* consists of 4 blocks, whose faces have different colors: some red, some blue, some white, and some green. None of the blocks has the same arrangement of colored faces. The object is to place them side-by-side in such a way that no two faces in front, on top, in back, or on the bottom are the same color.

CHAPTER 11

A SMORGASBORD OF PROBLEMS

This chapter presents problems covering topics in such areas as computer drawing, pattern recognition, mathematics, number system conversions, word and sentence generation, student testing, poetry generation, and statistics.

1. The computer can serve as a tool to draw pictures of Snoopy, Charlie Brown, or other simple figures. Write a program to draw a Christmas tree.

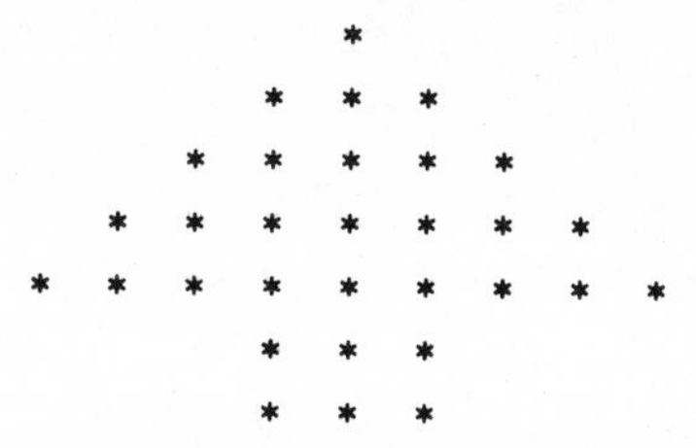

2. Produce a picture by printing X's on the output page. Some rules must be imposed concerning the way the X's are printed if the picture is to have some form.
3. Draw the face of a clock, including the 12 numbers and hands. Input to the program should be two numbers. These numbers will represent the time. For example, the number 7 and 30 would mean 7:30 o'clock.

4. Write a program that will draw a picture of the American flag. Use an asterisk to denote each star. Repeat each stripe by several lines repeated R's or W's, depending on the color of the stripe.
5. Write a program that will draw a picture of Snoopy, Charlie Brown, or some other cartoon figure. Draw the figure using X's.
6. Find all scores possible with 4 darts.

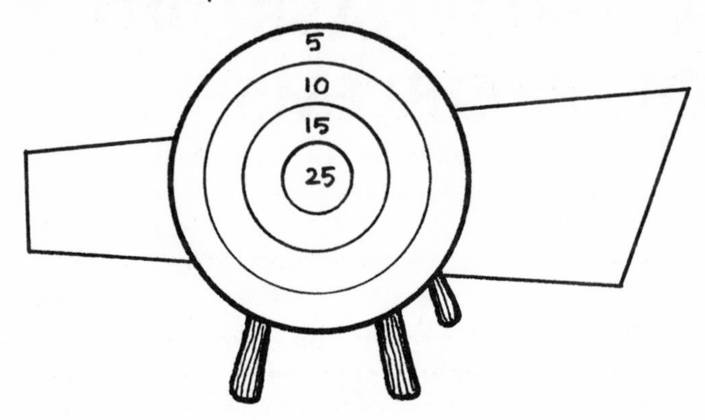

7. Write a program that generates random poetry.
8. From a data file of adjectives and nouns, generate random similes of the form (adjective) AS A (noun). For example, FAST AS A RABBIT, or RICH AS A MILLIONAIRE.
9. Write a program that generates random four-word sentences.
10. Write a program to generate random music.
11. Write a program that gives practice in finding the least common denominator (LCD) of two numbers.
12. Compute the fee of a baby-sitter for a given starting and ending time. Her sitting rate is 50 cents per hour until 11 P.M. and 75 cents per hour thereafter.
13. Prepare a program for converting units in cooking. There are 3 teaspoons to a tablespoon, 4 tablespoons to a quarter of a cup, and 2 cups to a pint. Using the code numbers 1 (teaspoon), 2 (tablespoon), 3 (cup), and 4 (pint) input the amount, the units used, and the new units desired. For example, the input 1, 4, 3 asks for the conversion of one pint to cups.

14. Write a program that will give you the correlation between college board scores and grade point averages for a group of students.

15. The town of Futureville had 21,609 residents in the year 2011. Every year there is a new baby born for each 210 residents and one death for every 263 residents. Every year exactly 61 new residents move to town and 84 move away. Determine the population in the year 2045.

16. Write a program to convert any number from 1 to 3000 to its equivalent Roman numeral. The seven Roman symbols are: M (1000); D (500); C (100); L (50); X (10); V (5); I (1). The rules for forming Roman numerals are: (a) If a symbol precedes one of smaller value, its value is added. (b) If a symbol precedes one of larger value, its value is subtracted; then the difference is added to the rest of the number. (c) Numbers are written as simply as possible using only C, X, and I as subtrahends. Some examples are MCMLXIV (1964), and DXLIX (549). Your program should accept as input the decimal number and output the Roman numeral. Convert the following numbers in your program: 1, 14, 400, 549, 999, 1964, 1984, 2500, 2994, 3000.

17. The United States currency system includes paper bills in denominations of $1, $2, $5, $10, $20, $50, $100, $500, $1000, $5000, and $10,000 and coins of 1, 5, 10, 25, 50, and 100 (silver dollar) cents. The $2 bill and 100 (silver dollar) coin are not included, due to their infrequent use in change making. Write a program to produce a printout based on a certain input value. For example, an input of 5367 ($53.67) would generate: fifty dollar bill, three one dollar bills, 1 fifty cent coin, 1 dime, 1 nickle, and 2 pennies.

18. Compute the batting average of the following five players and print a table of information in the order of decreasing batting average. The necessary data are given in the following table.

Player number	Times at bat	Hits
1	107	31
2	98	40
3	114	26
4	101	42
5	118	37

19. Players in a cricket team have batting figures for a season set out in a table as shown here:

Player's Number	Runs	Number of Innings	Times Not Out	Average
1	536	17	0	
2	642	14	2	
3	559	14	3	
.	.	.	.	
.	.	.	.	
.	.	.	.	
16	43	3	3	

Input the figures shown, calculate the batting averages, and print the complete table in descending order of averages.

20. Teach a computer to read by developing a standard set of characters somewhat as shown and writing a program by which the computer may recognize such patterns.

X				X
	X		X	
		X		
	X		X	
X				X

X

X	X	X		
X			X	
X				X
X			X	
X	X	X		

D

21. Compute the average number of times the letter"a" appears in all the words on one page of any book.

22. It has been speculated that if a group of monkeys were given typewriters and the monkeys hit the keys at random, over a long period of time they would eventually write every book that has been written including this one. Write a program to simulate a monkey typing. Assume that the events of the monkey hitting each key are independent. The procedure should stop when a monkey types the word APE.

23. Write a program that drills a student in performing simple arithmetic operations in addition, subtraction, multiplication, and division.

24. A girl in the fourth grade needs drill in adding fractions. Design a program to provide the drill, to keep score of her success, and to advance her to more difficult problems as she shows signs of mastering the given material.

25. Write a program that drills a student in adding rational numbers.

26. Write a program to quiz a student by having the program ask a question. If the student gets it right, go to the next question, but if he misses it, ask it again. If the student misses two or three times, go to the next question after giving him the right answer.

27. Northeastern Technical School offers both HONORS courses and REGULAR courses where letter grades A, B, C, D, and F are used to specify the student's achievement. To determine a student's grade point average the following weights are used:

	Point Value	
Letter Grade	HONORS	REGULAR
A	5	4
B	4	3
C	3	2
D	1	1
F	0	0

Write a program that will accept as input the *number* of REGULAR and HONORS letter grades received, and exhibit as output the student's grade point average.

28. Hungry, a small midwestern town, has 1,000 residents and is agriculturally self-sufficient (i.e., it cultivates enough food to feed itself). In fact, it produces enough food for 100,000 residents. However, every 10 years the population doubles and in that time enough food can be produced to feed 4,000 more people than in the previous 10 years. Output a table of data in the following format:

After Years	Population	Food Supply for
0	1,000	100,000
10	2,000	104,000
20	4,000	108,000
30	8,000	112,000

Have your program stop when the population exceeds the food supply.

29. Anyone who has recently flown in a jet aircraft is used to such statements as "The aircraft is now cruising at 35,000 feet," or "The ceiling is 60,000 feet." Though this information is certainly helpful, one may still want to know what the altitude is in miles. The conversion can be accomplished by dividing the altitude in feet by 5,280, the number of feet in a mile. Compute a table of converted values for 0 to 400,000 feet in steps of 20,000 feet.

30. Using standard U.S. coins, in how many ways can you make change for $1.00?

31. A part-time delivery boy is paid 1¢ the first day on the job, 2¢ the second day, 4¢ the third day, and so on, doubling each day on the job for 30 days. Calculate his wages on the 30th day and his total for the 30 days.

32. In an ecology course at Mainland High School, five examinations are given. Final grades for the course are to be based exclusively on examination scores, but the individual examinations are to be weighted differently in the computation. Scores received by six students are shown in the following table, which includes the weighting of each examination in parentheses following its number.

Student	Examination and weight				
	1 (.10)	2 (.15)	3 (.25)	4 (.15)	5 (.35)
1	63	68	72	89	93
2	99	100	76	83	94
3	53	68	63	75	78
4	93	97	100	89	91
5	75	72	81	78	84
6	78	81	69	75	79

Compute the final score for each student.

33. The average temperatures for six given months in Charleston, Texas follows.

June	78	December	41
July	89	January	36
August	93	February	27

Compute the average summer and winter temperatures in Charleston.

34. Given a knight located in column 5, row 4, print out all eight locations to which he is permitted to move by the rules of Chess.

35. The ABO blood groups in man are determined by a system of three alleles, A, B, and O. Genotypes AA and AO are group A; BB and BO are B; AB is group AB; and OO is group O. Write a program to determine whether the following proportions are consistent with the assumption of random mating. Given: 33.6 percent A, 20.8 percent B, 8.4 percent AB, and 29.3 percent O.

36. Seabreeze High School conducted a student survey to determine the most popular football player. The numbers 1, 2, 3, and 4 were used to determine the student votes.

    1 — vote for J. Superspeed, the halfback.
    2 — vote for S. Bigboy, the tackle.
    3 — vote for P. Smarts, the quarterback.
    4 — vote for B. Hands, the end.

    Fifty-two people participated in the survey and cast the following votes: 4, 1, 1, 2, 4, 1, 2, 3, 4, 4, 4, 1, 3, 3, 2, 4, 1, 2, 1, 4, 3, 3, 4, 1, 2, 4, 3, 2, 4, 4, 3, 1, 2, 4, 4, 3, 1, 1, 3, 4, 4, 4, 2, 1, 2, 4, 2, 4, 2, 1, 3, 4. Find out which football player was voted the most popular player.

37. A family is trying to determine the most economical nighttime setting for the thermostat in their house. The utility company estimates the cost, in cents, for each night at $(m - t)^2/10 + (72 - t)^2/100$ where m is the mean nighttime temperature and t is the thermostat setting, both measured in degrees Fahrenheit. The mean nighttime temperature m varies uniformly between 20°F and 70°F during the year. Simulate the values of m for 1 year and calculate the utility cost for a given value of t. Use the program with various values of t to find the most economical setting.

38. Three groups of rats are run in a skinner box bar-pressing for food reinforcements, under three shock levels. Write a program that will determine whether there is a significant effect of shock level on response rate.

39. 1729 is the first year that can be expressed in two different ways as the sum of two cubes. Write a program that will find two values for X and Y such that

$$X^3 + Y^3 = 1729$$

40. A duck and a pigeon are among the birds flying south for a winter vacation. The duck is 100 kilometers ahead of the pigeon and is flying at 95 kilometers per hour. The pigeon is flying along the same route at 105 kilometers per hour. How long will it take, in hours and kilometers, before the pigeon catches up to the duck?

NEW PROBLEMS

NEW PROBLEMS

NEW PROBLEMS